BEACH RESORT DESIGN
滨海度假村设计

《滨海度假村设计》编辑组 编　常文心 译

辽宁科学技术出版社
沈阳

图书在版编目 (CIP) 数据

滨海度假村设计 /《滨海度假村设计》编辑组编；
常文心译 . — 沈阳 : 辽宁科学技术出版社 , 2017.6
ISBN 978-7-5591-0128-0

Ⅰ . ①滨… Ⅱ . ①滨… ②常… Ⅲ . ①旅游度假村 –
建筑设计 – 案例 – 世界 Ⅳ . ① TU247.9

中国版本图书馆 CIP 数据核字 (2017) 第 072618 号

出版发行：辽宁科学技术出版社
　　　　　（地址：沈阳市和平区十一纬路 25 号　邮编：110003 ）
印 刷 者：辽宁新华印务有限公司
经 销 者：各地新华书店
幅面尺寸：245mm × 305mm
印　　张：52
插　　页：4
字　　数：250 千字
出版时间：2017 年 6 月第 1 版
印刷时间：2017 年 6 月第 1 次印刷
责任编辑：李　红
封面设计：李　莹
版式设计：李　莹
责任校对：周　文

书　　号：ISBN 978-7-5591-0128-0
定　　价：418.00 元

编辑电话：024-23280367
邮购热线：024-23284502
E-mail: 1207014086@qq.com
http://www.lnkj.com.cn

PREFACE
前言

THE ART OF DESIGNING LUXURY SEASIDE RESORTS
设计奢华滨海度假村的艺术

When embarking on the design of architecture, interiors or both of any resort, starting with a blank canvas and an integrative approach, emphasising the local context, client brief, environmental concerns, cultural sensitivities and desired end-user experience are the key. The surroundings alone of a seaside platform provide infinite possibilities for inspiration, as the springboard to award-winning, stellar design.

We try to create emotive, simple and elegant designs, for an integral experience suited to the surrounding environment, client requirements and lifestyles of potential guests of beach and seaside resorts. Attention to detail is witnessed in every space, where an appropriate ambience and mood are instilled in every shape. Clear and uncluttered concepts are infused with colour schemes and contrasts, through materials, surface accents and textures, to impart an air of confidence, comfort and welcome. All elements need to enjoy an aspirational harmony, in both the tangible and intangible experience, to ensure a complimentary overall dynamic to all areas. As the future moves swiftly towards more sustainable architecture and interior design solutions, with both clients and operators requesting greener options, the outlook for seaside resorts is a bright, environmentally-friendly driven quest. Utilising locally-sourced materials, suiting the design to the geographic location, providing energy- and light-saving alternatives, maximising natural daylight and views, as well as preserving the natural habitat, are all firmly entrenched in the design ethos. It is only a matter of time, as a natural progression, before these ideals and more become securely engrained in the basic underlying structure for approaching all seaside resort design to come.

Scott Whittaker
Group Creative Design Director
dwp | design worldwide partnership

在度假村的建筑和设计中，从一张白纸入手，进行综合设计，设计师的重点始终是突出当地环境、客户需求、环境因素、文化敏感度和终端用户体验。滨海平台周边的环境为设计提供了源源不断的灵感、是优秀设计的跳板。

我们尽量打造感性、简洁而优雅的设计，打造适合滨海度假村周边环境、客户需求以及潜在用户生活方式的综合设计。度假村的每个角落都着重于细节，力求在各处都营造出合适的气氛和心情。简洁明晰的概念与色彩搭配和对比通过材料、特色装饰和材质结合在一起，为度假村注入了自信、舒适和宾至如归的感觉。所有元素在有形和无形的体验中都显得无比和谐，保证了所有区域都活力十足。

未来正向可持续建筑和室内设计方面飞速发展，客户和运营商都要求更加绿色的设计。因此，滨海度假村的前景一片光明，不断追求环保。设计始终以采用本地材料、适应地理位置、提供节能方案、最大化自然采光和视野以及保护自然栖息地为理念。实现这些理想并将以上理念植入滨海度假村的基础结构中只是一个时间问题。

斯科特·维泰克
团队创意设计总监
dwp |全球设计事务所

CONTENTS
目录

Anantara Kihavah Villas

安娜塔拉基阿瓦别墅度假村

Completion date: 2011
Location: Baa Atoll, Maldives
Designer: Group X Design Associates (Architect and Structure Design)
August Design (Interior Design for Guest Rooms and Public Areas)
Poole Associates (Interior Design for Food & Beverage Outlets, Lobby and Library)
Abacus Design (Interior Design for Spa)
Photographer: Anantara Hotels, Resorts & Spas
Area: 120,000 sqm

竣工时间：2011年
项目地点：马尔代夫，芭环礁
设计师：Group X Design Associates 建筑师事务所（建筑与结构设计）
August Design 设计事务所（客房与公共空间室内设计）
Poole Associates 设计事务所（餐饮，大堂与图书室室内设计）
Abacus Design 设计事务所（spa室内设计）
摄影师：安娜塔拉酒店与水疗度假村集团
项目面积：120,000平方米

Anantara Kihavah Villas is surrounded by the jewel-coloured waters and uninhabited islands of the majestic Indian Ocean. It offers intimate luxury and total surrender to the ambiance and peacefulness of the greatest paradise on earth. 78 spacious private pool sanctuaries, ranging from 260 to 2,730 square metres, are either poised over the water with sweeping ocean views or nestled along a pristine stretch of private beach. Comprising 74 one-bedroom villas, 3 two-bedroom residences and a three-bedroom residence, each hideaway boasts large personal infinity-edge pools, dining pavilions, expansive wooden sundecks and ample lounging areas complete

with swinging daybeds, hammocks and sun loungers. All villas feature his and hers walk-in wardrobes, rain showers, outdoor showers and over-sized bathtubs big enough for two. Sunken glass bottom bathtubs in the over water villas and residences offer a mesmerizing view of the sparkling ocean, both below and beyond the adjacent infinity-edge pool.

Seamless style is achieved with contemporary interiors featuring local accents in every room and colour schemes that take inspiration from the surrounding coral reefs. Guest villas are at one with the natural tropical surroundings with coconut trees jutting through some of the luxury villas' natural wood floors and lush canopies offering shade and seclusion. The residences additionally include an expansive indoor living and dining room, pantry and guest powder room, while the three-bedroom residence boasts a jungle massage pavilion, private beach cabana and a barbeque station.

There are six restaurants and bars, each offering a distinctive setting, menu and ambiance that divulges a different view of the world's most envied utopia. Highlighting the resort's idyllic location in the dazzling Indian Ocean is the exclusive "Sea. Fire. Salt. Sky". This unique under and over water gastronomic concept

1. Beach Pool Villa
2. Two-Bedroom Beach Pool Residence
3. Three-Bedroom Presidential Beach Residence
4. Over-Water Pool Villa
5. Two-Bedroom Over-Water Pool Residence
6. Plates-Breakfast & Buffet Dinner
7. Manzaru-Pool Bar and Restaurant
8. Reception
9. Library and Boutique
10. Specialty Restaurant
11. Anantara Spa
12. Thiththi Boli Kids Club
13. Recreation Centre
14. Dive & Watersports Centre
    Conference Centre
15. Tennis Court

1. 海滨泳池别墅
2. 两室海滨泳池别墅
3. 三室总统海滨别墅
4. 水上泳池别墅
5. 两室水上泳池别墅
6. Plates餐厅—早餐与自助餐
7. Manzaru餐厅—泳池吧与餐厅
8. 前台
9. 图书室和精品店
10. 特色餐厅
11. 安纳塔拉水疗中心
12. Thiththi Boli儿童俱乐部
13. 娱乐中心
14. 潜水&水上运动中心
    会议中心
15. 网球场

1. Resort plan
2. Anantara Spa

1. 度假村平面图
2. 安娜塔拉水疗中心

features four remarkable venues, each offering a different type of cuisine that is as unique as the perspective. In the evening the upstairs restaurant, "Manzaru", transforms into a leisurely Italian delight offering regional specialties. "Dining by Design" serves at a table for two in dreamy locations, from an isolated sandbank in the Indian Ocean to the deck of a private yacht.

The Anantara Spa offers a therapeutic range that combines traditional Asian therapies and luxuriously pampering Elemis products. The spa comprises of six over-water treatment suites, a beauty salon, hair and nail salon and two single facial suites accompanied by a relaxation area with large sweeping decks and a Jacuzzi in between two oversized plunge pools that are suspended over the ocean.

1. Dining by Design
2. Two-Bedroom Over-Water Pool Residence lounge
3. Library

1. 设计就餐区
2. 两居室水上泳池别墅休息室
3. 图书室

安娜塔拉基阿瓦别墅度假村四面环绕着宝石般色彩的海水和浩瀚的印度洋上的无人岛。该度假村拥有无比奢华的亲密氛围，堪称是静谧优雅、轻松愉快的人间天堂。78个宽敞的私人泳池圣殿，面积260平方米到2730平方米不等，有的泰然自若地挺立于水面之上，一览壮观的海景；有的排列在纯朴的私人海滩沿岸。安娜塔拉度假村共有74个一居室别墅和4个两居室住宅，每栋建筑都拥有大型私人无边际泳池、餐饮场馆、广阔的木制阳台和宽敞的休息室，并配备舒适的沙发床、吊床和日光浴浴床。所有别墅都拥有步入式衣柜、淋浴花洒、露天淋浴和巨大的双人浴缸。下沉式玻璃底水上浴缸为人们提供了无边界泳池下方和旁边闪闪发光的海洋美景。

富有当地风情的现代室内设计遍布每个房间，而色彩搭配则从周边的珊瑚礁中获得了灵感。别墅通过奢华的实木地板和丰富的屋顶遮阳天篷与热带环境中的椰子树相互呼应。别墅拥有宽敞的室内起居室和餐厅、备餐间和客用化妆间。三卧别墅还享有丛林按摩亭、私人海滩小屋以及烧烤平台。

度假村共有六家餐厅和酒吧，每家都拥有独一无二的布景、菜单和环境，打造了世上最令人艳羡的乌托邦。餐厅以璀璨印度洋之上的田园风情为特色，拥有独一无二的"海洋，火焰，盐，蓝天"。这种水上和水下的美食概念以四种非凡的元素为特色，提供了与景观同等别致的美食。晚上，楼上餐厅曼扎鲁会变成一家休闲意大利美食坊，专门提供当地特色美食。设计餐厅将提供双人梦幻就餐地点，从印度洋上的独立沙洲到私人游艇的甲板上，十分私密。

安娜塔拉水疗中心所提供的各种疗法结合了传统亚洲疗法和奢侈贴心的埃勒米斯产品。水疗中心由六间水上治疗套房、美容沙龙、美发美甲沙龙、两间担任面部护理套房以及休闲区组成。休闲区配有巨大的平台，两个巨大的跌水泳池漂浮在印度洋上，二者之间设有舒适的按摩浴缸。

1. Manzaru Restaurant long table
2. Manzaru Bar

1. 曼扎鲁餐厅长桌
2. 曼扎鲁酒吧

1. Swimming pool
2. Two-Bedroom Over-Water Pool Residence and swimming pool
3. Beach Pool Villa
4. Beach Pool Villa swimming pool

1. 游泳池
2. 两居室水上泳池别墅和游泳池
3. 海滩泳池别墅
4. 海滩泳池别墅游泳池

1. Beach Pool Villa romantic bathtub
2. Over-Water Pool Villa bedroom
3. Beach Pool Villa interior

1. 海滩泳池别墅的浪漫浴缸
2. 水上泳池别墅卧室
3. 海滩泳池别墅室内

Velassaru Maldives

马尔代夫维拉沙鲁岛度假村

Completion date: 2009
Location: Velassaru Island, Maldives
Designer: The Olive Garden Singapore Design Firm
Photographer: Velassaru Maldives
Area: 8,000 sqm

竣工时间：2009年
项目地点：马尔代夫，维拉沙鲁岛
设计师：新加坡橄榄园设计事务所
摄影师：马尔代夫维拉沙鲁岛度假村
项目面积：8,000平方米

With a huge white sandy beach surrounding the island, Velassaru Maldives is nestled in a grand blue lagoon. It offers seven types of accommodations. The resort has 52 Deluxe Bungalows which is only a few steps away from the beach surrounded by lush tropical foliage, 30 Beach Villas of elegantly designed interiors, 17 Water Bungalows with over-water decks and sun lounges, 24 Water Villas perched over the azure blue lagoon, and four premium Water Villas that come with a pool. Velassaru also offers a Pool Villa located in the very heart of the island surrounded by flourishing vegetation, and a Velassaru Water Suite of complete privacy with an oversized plunge pool overlooking the ocean.

The resort has seven dining facilities that offer world cuisines and cocktails. They also provide a grand collection of unique private dining experiences. The resort has five restaurants, Etesian (a journey of Mediterranean flair), Sand, Teppanyaki, Vela, and Turquoise. Fen Bar, the main bar, and Chill Bar offer a wide variety of alcoholic and non-alcoholic beverages and healthy drinks.

This resort also has a PADI Certified Dive School and marine discovery activities. The Glide Water Sports Centre at Velassaru also offers many water activities. For other activities the resort also offers a fully equipped gymnasium and a tennis court.

Velassaru also offers a spa of solitude, space and peace, with ten treatment rooms, designed to refresh and rejuvenate your spirit as well as your physical being. It gives you privacy and tranquility and allows you to turn away from the burdens of the world while you experience the invigorating treatments.

This resort also allows you to experience an inner journey of positiveness and happiness, with their yoga instructor. It allows you to develop the unitive disciplines of relaxation and flexibility. The yoga at Velassaru indeed relieves stress and cultivates a quiet and peaceful mind, more capable of alertness and concentration.

1. Immersion Dive Centre
2. Reception
3. Fen Bar
4. Safani Jewellery Shop
5. Boutique
6. Turtiquoise
7. Pool Deck Pizzeria
8. Vela Restaurant
9. Sand Restaurant
10. Teppanyaki Restaurant
11. Etesian Restaurant
12. Glide Water Sports Centre
13. Water Villas and Bungalows
14. Energy Fitness Centre
15. Chill Bar
16. Tennis Court

1. 潜水中心
2. 前台
3. 沼泽吧
4. 萨法尼珠宝店
5. 精品店
6. 特尔蒂克餐厅
7. 泳池甲板比萨店
8. 维拉餐厅
9. 沙滩餐厅
10. 铁板烧餐厅
11. 季风餐厅
12. 滑水运动中心
13. 水上别墅
14. 能量健身中心
15. 冷吧
16. 网球场

马尔代夫维拉沙鲁岛度假村位于一个巨大的蓝色环礁湖中，周围环绕广阔的白色沙滩。马尔代夫维拉沙鲁岛度假村拥有7种不同类别的客房，其中52栋豪华屋住宅距离翠绿热带植物包围的海滩仅几步之遥，30栋海滩别墅拥有精致优雅的室内设计，17栋水上屋住宅配备水上露台和日光浴室，24栋水上别墅栖息于蔚蓝色的环礁湖上，4栋高级水上别墅都配备一个游泳池。马尔代夫维拉沙鲁岛度假村还拥有一栋泳池别墅和一栋维拉沙鲁岛水上别墅，泳池别墅位于岛屿中央，四周苍翠植被环绕；维拉沙鲁岛水上别墅四周环境隐蔽，拥有一个超大型跌水潭，游客可在此一览海洋的壮观景色。

马尔代夫维拉沙鲁岛度假村中的7种餐饮设施提供世界顶级的美食与鸡尾酒，还提供一系列别致的私人用餐体验。该度假村共有5家餐厅：带您领略地中海风情的季风餐厅、沙滩餐厅、铁板烧餐厅、维拉餐厅和特尔蒂克餐厅。沼泽吧是主酒吧，而冷吧则提供种类繁多的酒品、无酒精饮料和保健饮品。

马尔代夫维拉沙鲁岛度假村还拥有一所经过专业潜水教练协会认证的潜水学校和一个生物安全实验室海洋探索中心。滑水运动中心还提供多种水上运动。该度假村还拥有一个设施配备齐全的体育馆和一个网球场，可进行其他活动。

马尔代夫维拉沙鲁岛度假村还拥有一个广阔、安静、隐蔽的水疗中心，并配备10间理疗室，游客可在此放松身心，体验幽静的隐私空间，让您远离尘世的压力，享受无比舒适的理疗服务。

马尔代夫维拉沙鲁岛度假村还有专业的瑜伽教练带您体验内心世界的快乐，重获对生活的信心。您可跟随瑜伽教练进行全身的舒展训练和灵活度训练。马尔代夫维拉沙鲁岛度假村的瑜伽训练会让您真正地释放压力，拥有安静祥和的内心，从而更好地集中精力、保持清醒的头脑。

1. Beach Villa exterior
2. Resort plan
3. Sand Restaurant

1. 海滨别墅外景
2. 度假村平面图
3. 沙滩餐厅

1. Pool
2. Pool Villa poolside
3. Water Suite pool deck
4. Water Villa deck

1. 泳池
2. 泳池别墅泳池边
3. 水上套房泳池露台
4. 水上别墅露台

1

1. Pool Villa pool and relax area
2. Beach Villa bathroom

1. 泳池别墅的游泳池和休息区
2. 海滨别墅浴室

1. Etesian wine cellar
2. Cosy sofa at Fen Bar
3. Reception
4. Water Villa overview

1. 季风酒窖餐厅
2. 沼泽吧内舒适的沙发
3. 接待大厅
4. 水上别墅全景

1. Pool Villa living area
2. Water Suite room
3. Water Bungalow
4. Deluxe Bungalow
5. Water Suite bathroom

1. 泳池别墅起居室
2. 水上套房卧室
3. 水上别墅卧室
4. 顶级别墅
5. 水上套房浴室

Naladhu, Maldives

马尔代夫娜拉杜岛度假村

Completion date: 2007
Location: South Male Atoll, Maldives
Designer: Mohamed Shafeeq, Group X Design Associates
Photographer: Naladhu, Maldives
Area: 22,140 sqm

竣工时间：2007年
项目地点：马尔代夫，南阿累环礁
设计师：Mohamed Shafeeq、Group X Design
Associates建筑师事务所
摄影师：马尔代夫娜拉杜岛度假村
项目面积：22,140平方米

Located in the aquatic setting of the Maldives, one of the world's most celebrated tropical havens, Naladhu, Maldives offers a unique lifestyle to those who appreciate the sophistication and charm of times past. Coupling the ingenious talents of Maldivian architect Mohammed Shafeeq and Thailand-based interior designer Julian Coombs, Naladhu has captured the elegant aesthetics of a slightly colonial era, with hints of Sri Lankan regality in its design.

Naladhu means "pretty little island". Situated on a private island ringed by coral gardens and vibrant marine life, where the stunning turquoise tones of the Indian Ocean

are complemented by the exotic scent of nearby palm trees, Naladhu promises an unrivalled experience to those who savour the finer aspects of life.

Each elite dwelling is a colossal space 330 square metres in size and punctuated by a lofty white-gabled roof, expansive glass-panelled doors and generous living, dining and bedroom spaces. While the historic elements are intrinsic to Naladhu's atmosphere, each detail of the Naladhu house has been conceived to serve every guest's whim. A personal plunge pool, landscaped garden and vast teak sundeck make for an exotic experience within the privacy of the enclosure.

Bungalows each have a voluminous terrazzo bath tub with free-formed edges as well as an open-faced shower which allows for a view from within but remains hidden to anyone outside the house enclosure. The objective of Naladhu is to provide and pamper guests with unsurpassed quality and service.

1

1. Reception
2. Fitness Centre
3. Living Room
4. Swimming Pool
5. Bungalows

1. 前台接待处
2. 健身中心
3. 客厅
4. 游泳池
5. 别墅

马尔代夫娜拉杜岛度假村地处世界上最著名的热带天堂——马尔代夫的水生环境中。马尔代夫娜拉杜岛度假村为那些崇尚古典时光魅力与精湛技艺的人士提供了别致的生活方式。马尔代夫娜拉杜岛度假村结合了马尔代夫本土设计师Mohammed Shafeeq和泰国室内设计师Julian Coombs高超的设计技巧，成功地捕捉到了略带殖民地色彩的典雅唯美，设计体现了斯里兰卡王室的奢华。

娜拉杜的意思是"优美的小岛"。 马尔代夫娜拉杜岛度假村坐落在一个四周布满珊瑚花园、海洋生物活跃的私人小岛上，成排的棕榈树点缀着印度洋碧玉般的海水。娜拉杜岛度假村为那些追求优质生活、高品位生活的人士提供了一次无与伦比的度假体验。

马尔代夫娜拉杜岛度假村中每一栋精致的别墅都拥有330平方米的超大面积，采用高高的白色人字形屋顶和宽大的玻璃格子门，每栋别墅都拥有宽敞的客厅、餐厅和卧室。娜拉杜岛度假村不仅拥有特色的古典元素，甚至房屋的每一处细节设计都满足了游客的奇妙幻想。

每个小木屋都配备一个宽大的水磨石浴缸，浴缸的边缘采用不规则设计；还有一个单片淋浴器，游客可以边淋浴，边观望室外景观，淋浴器安放在十分隐蔽的地方，任何人都无法看到。为顾客提供非凡卓越的品质与服务一直是娜拉杜岛度假村的宗旨。

2

1. Resort plan
2. Poolside

1. 度假村平面图
2. 游泳池边

1. Living room deck
2. Morning yoga by the pool

1. 客厅平台
2. 泳池边的晨间瑜伽

1. Naladhu Beach House
2. Your sala and pool
3. Naladhu Ocean House

1. 娜拉杜海滨别墅
2. 大厅和游泳池
3. 娜拉杜海洋别墅

1. Naladhu reception
2. Beach House bedroom
3. Beach House bedroom

1. 娜拉杜前台
2. 海洋别墅卧室
3. 海洋别墅卧室

Centara Grand Island Resort & Spa Maldives

马尔代夫中央格兰德岛度假村

Completion date: 2009
Location: South Ari Atoll, Maldives
Designer: Miaja Design Group
Photographer: Centara Grand Island Resort
& Spa Maldives
Area: 46,800 sqm

竣工时间：2009年
项目地点：马尔代夫，南阿里环礁
设计师：米亚加设计集团
摄影师：马尔代夫中央格兰德岛度假村
项目面积：46,800平方米

Set amongst the perfect islands and blue ocean of the South Ari Atoll (Alifu Atoll) in the Republic of Maldives, featuring 112 suites and villas, the resort offers outstanding diving opportunities including an excellent house reef complete with dedicated sunken ship wreck and is within easy reach of some of the top dive spots in the Maldives. Centara Grand Island Resort & Spa Maldives offers more than simply daytime sun, sea, and sand. The resort features a variety of dining options, pools, recreation activities, sports and fitness facilities, bars and lounges with nightly entertainment, SPA Cenvaree, and a Kids Club.

Centara Grand Island Resort & Spa Maldives offers beach front and over-water accommodations divided into seven categories which range from 87 to 159 square metres complete with a "resort within a resort" accommodation experience – Island Club. The Island Club concept offers premium guests exclusive use of the island clubhouse, swimming pool, and a more expansive choice of beverage using popular branded spirits. Additionally, the club pool features refreshing Spa Cenvaree mist sprays.

Forty-two Beach Suites accommodate up to three adults or two adults and two children per room. This spacious beach front suite comprises 87 square metres separated into two floors and connected via staircase and offers a master king or twin bedroom complete with a separate furnished living room, bathroom with Island bathtub and heavenly rain shower.

Ten Deluxe Water Villas and eight Deluxe Water Villas – Island Club respectively accommodate up to three adults or two adults and one child per room. The spacious over-water villa comprises 87 square metres and offers a master king or twin bedroom complete with furnished living room area, bathroom with personal jacuzzi spa and heavenly rain shower.

Ten Deluxe Family Water Villas accommodate up to

1. Luxury Sunset Water Villas

2. Beach Suites

3. Deluxe Family Water Villas

4. Luxury Beach Front Pool Villas

5. Deluxe Water Villas

6. Island Club

7. Azzuri Mare (Italian Restaurant)

8. Arrival Pavilion, Reception Lounge & Aqua Bar

9. Water Sports Centre

10. Lotus (Thai Restaurant)

11. Spa Cenvaree

12. Rock Climbing

13. Dive School & Gym

14. Coral Bar

15. Reef Restaurant

1. 奢华落日水上别墅   13. 潜水学校&健身房

2. 海滨套房   14. 珊瑚吧

3. 豪华家庭水上别墅   15. 暗礁餐厅

4. 奢华海滨泳池别墅

5. 豪华水上别墅

6. 小岛俱乐部

7. 阿祖里迈尔餐厅（意大利餐厅）

8. 到达大厅、接待休息室和水吧

9. 水上运动中心

10. 莲花餐厅（泰国餐厅）

11. 桑瓦里水疗馆

12. 攀岩活动

1. Infinity pool
2. Resort plan
3. Island Club

1. 无边界游泳池
2. 度假村平面图
3. 小岛俱乐部

three adults or two adults and two children per room. This spacious over-water villa comprises 93 square metres and offers a master king bedroom complete with furnished living room area, bunk beds in a separate kids area, bathroom with personal jacuzzi spa and heavenly rain shower.

Twenty Luxury Sunset Water Villas accommodate up to three adults or two adults and 1 child per room. This spacious over-water villa comprises 110 square metres separated into two floors and connected via outdoor staircase. Extensive outdoor living area is located over two floors. The upper deck is outdoor with a lounging and relaxing area offering panoramic ocean views.

Fourteen Luxury Sunset Water Villas – Island Club accommodate up to three adults or two adults and one child per room. This spacious over-water villa comprises 110 square metres including two-storey sun deck connected via outdoor staircase.

Eight Two-bedroom Luxury Beach Front Pool Villas – Island Club accommodate up to four adults or three adults and two children or two adults and three children per room. This spacious beach front villa comprises 159 square metres and offers two bedrooms (king and twin bedroom) complete with a separate furnished living room, exclusive 22-square-metre private pool, bathroom with Island bathtub and heavenly rain shower.

马尔代夫中央格兰德岛度假村在南阿里环礁的完美岛屿和碧蓝海水环绕之中，拥有112套套房和别墅，并且为游客们提供了非凡的潜水机会。沉船让卓越的水下环游之旅更加完美，同时度假村还紧邻马尔代夫最顶级的潜水景点。马尔代夫中央格兰德岛度假村不仅仅是阳光、海洋和沙滩，还以一系列特色服务来吸引游客——美味的餐厅、游泳池、各种各样的娱乐活动、运动健身设施、让夜生活丰富多彩的酒吧和休息室、桑瓦里水疗馆和儿童俱乐部，都为游客提供了非凡的体验。

马尔代夫中央格兰德岛度假村所提供的海滩住宿和水上住宿共分为七个种类，面积从87平方米到159平方米不一，并且还设有一个"度假村中的度假村"——小岛俱乐部。小岛俱乐部为顶级宾客提供专属会所、游泳池和更多的高档酒水选择。此外，俱乐部的游泳池以清爽的水疗喷雾为特色。

在42套海滩套房中，每个房间可容纳到2-3名成人和两名儿童。宽敞的海滩套房总面积为87平方米，分为两层，通过楼梯相连，设有一间巨大的主卧或两间卧室、独立的起居室和浴室（带有小岛浴缸和天堂淋浴）。

在10套豪华水上别墅和8套小岛俱乐部专属豪华水上别墅中，每个房间可容纳2-3名成人和一名儿童。水上别墅总面积87平方米，设有一间巨大的主卧或两间卧室、独立的起居室和浴室（带有私人极可意按摩水疗和天堂淋浴）。

在10套豪华水上别墅中，每个房间可容纳2-3名成人和两名儿童。水上别墅总面积93平方米，设有一间巨大的主卧、独立的起居室、儿童房双层床和浴室（带有私人极可意按摩水疗和天堂淋浴）。

在20套奢华水上别墅中，每个房间可容纳2-3名成人和一名儿童。水上别墅总面积110平方米，分为上下两层，通过露天楼梯相连。宽阔的露天起居区设在屋顶上。露天上层平台上设有休闲区，提供了完美的海洋全景。

在14套奢华水上别墅——小岛俱乐部中，每个房间可容纳2-3名成人和1名儿童。水上别墅总面积110平方米，分为上下两层，通过露天楼梯相连。

在8套双卧室奢华海滨泳池别墅——小岛俱乐部中，每个房间可容纳3-4名成人和两名儿童，或者3名成人和3名儿童。海滨泳池别墅总面积159平方米，设有两间卧室（国王卧室与双人卧室）、独立的起居室，22平方米的私人泳池，还有带小岛浴缸和天堂淋浴的浴室。

1. Reef deck
2. Azzuri Mare seafood restaurant
3. Aqua over-water bar with snacks and drinks
4. Coral Bar

1. 暗礁平台
2. 阿祖利马尔海鲜餐厅
3. 提供点心和饮品的水上酒吧
4. 珊瑚吧

1. Reception
2. Reef restaurant
3. Deluxe Water Villa
4. Island Club

1. 前台
2. 暗礁餐厅
3. 豪华水上别墅
4. 小岛俱乐部

1. Luxury Sunset Water Villa
2. Luxury Sunset Water Villa
3. Deluxe Water Villa
4. Spa Cenvaree

1. 奢华落日水上别墅
2. 奢华落日水上别墅
3. 豪华水上别墅
4. 桑瓦里水疗中心

Halaveli Resort

哈拉薇丽度假村

Completion date: 2009
Location: Alifu Alifu Atoll, Maldives
Designer: Martin Branner
Architect: LGX Associates
Photographer: Martin Branner
Area: 125,000 sqm

竣工时间：2009年
项目地点：马尔代夫，阿里夫环礁
设计师：马丁·布兰那
建筑师：LGX建筑事务所
摄影师：马丁·布兰那
项目面积：125,000平方米

Floating in the North Ari atoll and shaped like a curved Dhoni (Maldivian boat) is the five-star Constance Halaveli Resort. It is a place where time seems to have stopped and dreams become reality. Water and sand combine and lie in contrast to the exuberant green of the foliage.

The design concept of the resort is conceived as a presentation of luxury and exquisite craftsmanship. The design scheme appeals to international travellers who desire to have an undisturbed and recreational time at the beach, highlighting exceptional guest services and amenities, outstanding views, and cutting-edge technologies. Amenities designed using natural materials,

fabrics and furnishings offer a sense of welcome, while protecting the environment at the same time.

The shadow of the 86 villas falls on the turquoise lagoon. There are 57 Water Villas of 100 square metres with private plunge pool & sun deck, 20 Beach Villas of 350 square metres with private plunge pool & garden, 8 Double Storey Beach Villas of 410 square metres with private plunge pool & garden, and 1 Presidential Beach Villa of 700 square metres with private plunge pool. All the villas are air-conditioned and feature their own private plunge pool located on the beach or over the water. They are comfortably furnished using modern wood and marble. Each villa has also a furnished terrace or balcony and bathroom comprising of separate shower (inside or outside), bath/WC and complete but unique Spa Experience offering natural healing and nurturing beauty from within.

The three restaurants and the spa ensure that both body and mind are well cared for. There's also another considerate design for children. The "Kuda Club" welcomes children aged from 4 to 12 years. Open-air and indoors entertaining as well as educational activities are specially designed for kids. Constance Halaveli Resort is a place to relax and regenerate in overwhelming peacefulness.

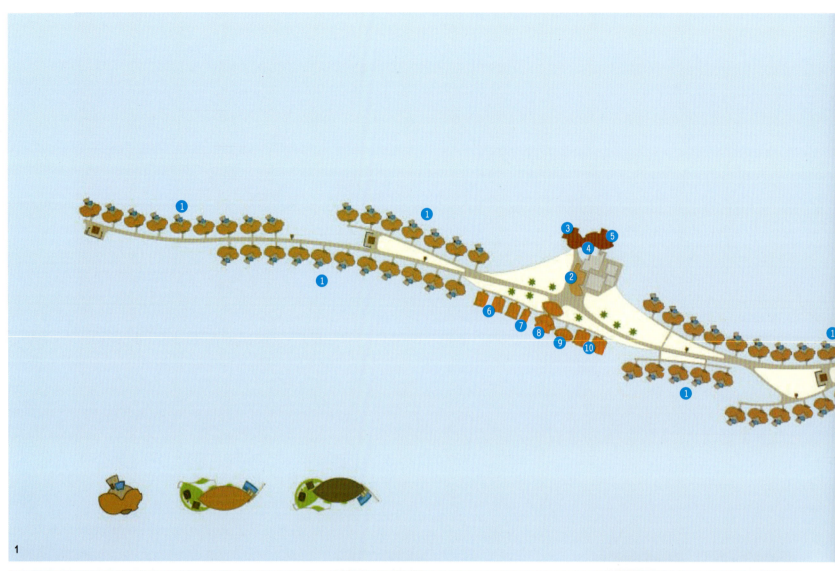

1

2

1. Water Villa
2. Water Villa Reception
3. Speciality Bar: JING
4. Wine Cellar
5. Speciality Restaurant: JING
6. Single Treatment Rooms
7. Ayurveda
8. Spa Reception
9. Manicure & Pedicure
10. Double Treatment Rooms
11. Beach Villa
12. Double Storey Beach Villa
13. Main Jetty
14. Dive Centre
15. Main Reception
16. Boutiques
17. Water Sports Centre
18. Kids Club
19. Main Bar: JAHAZ
20. Main Pool
21. Gym
22. Mosque
23. Presidential Beach Villa
24. Beach Grill: MEERU
25. Main Restaurant: JAHAZ
26. Tennis Court

1. 水上别墅
2. 水上别墅接待处
3. 特色酒吧：晶
4. 酒窖
5. 特色餐厅：晶
6. 单间治疗室
7. 阿育吠陀治疗室
8. 水疗中心前台
9. 美甲
10. 双人治疗室
11. 海滨别墅
12. 双层海滨别墅
13. 主码头
14. 潜水中心
15. 主接待处
16. 精品店
17. 水上运动中心
18. 儿童俱乐部
19. 主酒吧：雅杂
20. 主游泳池
21. 健身房
22. 清真寺
23. 总统海滨别墅
24. 海滨烧烤店：美禄
25. 主餐厅：雅杂
26. 网球场

1. Resort plan
2. Beach view

1. 度假村平面图
2. 海滩景色

五星级的康斯坦士哈拉薇丽度假村漂浮在北阿里夫环礁之上，宛如一条弯曲的多尼船（马尔代夫特色小船）。在这里，时间似乎停止，梦想变成现实。水和沙相互融合，与茂密的植物形成了鲜明对比。

度假村的设计理念尽显奢华和精致的手工艺。设计专门吸引渴望在沙滩上享受无忧无虑的休闲时光的国际游客，以非凡的游客服务和娱乐设施、出色的景点和尖端的技术为特色。采用天然材料、织物和装饰进行设计的娱乐设施热情洋溢，同时又具有环保价值。

86座别墅倒映在绿松石般的环礁湖里。其中包括57座配有私人游泳池和日光浴平台的水上别墅（100平方米）、20座配有私人游泳池和花园的海滨别墅（350平方米）、8座配有私人游泳池和花园的双层海滩别墅（410平方米）和一座配有游泳池的总统海滩别墅（700平方米）。全部别墅都配有空调设施，私人游泳池设在海滩或者水上。每座别墅都有一个装修的平台或阳台。浴室由独立淋浴间（设在室内或室外）、洗手间和完整的温泉设施组成，为游客提供自然疗法并塑造由内而外的美。

三家餐厅和温泉设施保证了游客的身心都得到很好的照料。度假村还为儿童提供了贴心的设计。"库达俱乐部"欢迎4-12岁的儿童前来游玩。露天和室内游乐设施以及教育活动都专为儿童设计。康斯坦士哈拉薇丽度假村是一个宁静的休闲、新生之所。

1. Main entrance
2. JING Restaurant
3. JAHAZ Restaurant
4. Beach villa exterior

1. 正门入口
2. "晶" 餐厅
3. "雅杂" 餐厅
4. 海滨别墅外景

1. Spa reception
2. Treatment room
3. Kids club
4. Gym

1. Spa接待处
2. 理疗室
3. 儿童俱乐部
4. 健身房

1. Presidential villa bedroom
2. Water villa bedroom
3. Beach villa living room
4. Beach villa bedroom

1. 总统套房卧室
2. 水上别墅卧室
3. 海滨别墅起居室
4. 海滨别墅卧室

Karma Kandara

卡玛坎达拉度假村

Completion date: 2008
Location: Bali, Indonesia
Designer: Grounds Kent
Photographer: Gabriel Ulung Wicaksono
Area: 45,000 sqm (not including 800 sqm for Nammos Beach Club)

竣工时间：2008年
项目地点：印度尼西亚，巴厘岛
设计师：格朗兹·肯特建筑师事务所
摄影师：加布里埃尔·犹朗·维卡索诺
项目面积：45,000平方米（不包括800平方米的纳莫斯海滩俱乐部）

Karma Kandara occupies a spectacular cliff top high above the Indian Ocean at the very tip of Bali's elevated southern peninsula. Bridges, stone walkways and little paths meander through vivid tropical gardens, connecting 46 expansive pool residences. Both the destination restaurant with its rooftop bar and the spa are suspended on rocky outcrops, while a private inclinator takes guests down to Nammos Beach Club, set at the base of the cliff. All Karma Kandara's 2, 3 & 4 bedroom residences are contemporary elaborations on the Balinese compound style, comprising two or three pavilions with roofs of alang grass or sirap wood tiles, framing a central lap pool

and a courtyard. Living rooms have an airy, expansive feel with their high-pitched ceilings and creamy marble floors. Décor evokes an uncluttered Mediterranean ambience, with richly upholstered furnishings in shades of chocolate, cinnamon, copper, bronze and amber, accented with local artworks and antique artifacts. All bedrooms boast en-suite semi-outdoor bathrooms with grandiose stand-alone tubs. The Grand Residence is at the very brink of the cliff top with its horizon pool seemingly spilling into the sea.

Suspended 85 metres above the surf on a rocky outcrop, Di Mare Restaurant designed in the manner of a floating viewing platform is one of the most dramatically situated destination restaurants in Bali. Set on the rooftop of Di Mare Restaurant, Temple Lounge evokes a North African ambience. Plush, colourful banquettes, Shisa pipes and giant Moroccan stained glass lanterns create a Kasbah-like setting. Nammos Beach Club is a private folly that conjures the romance of the Greek Islands. The Karma Spa is perched dramatically on a shoulder of rock overlooking the ocean.

1. The villas
2. Nammos Beach Club

1. 别墅
2. 纳莫斯海滩俱乐部

卡玛坎达拉度假村位于印度洋上一个陡峭壮观的悬崖上方，地处巴厘岛高耸的南部半岛的最顶端。卡玛坎达拉度假村拥有46个广阔的泳池别墅，穿梭于热带园林中的小桥、石质通道和小路将各个别墅相连。无论是带屋顶酒吧的招牌餐厅还是水疗中心，都悬浮于岩石的上方，而一个私人观光电梯则将顾客带到了下面的纳莫斯海滩俱乐部，海滩俱乐部位于悬崖的底部。

卡玛坎达拉度假村拥有两居室、三居室和四居室等不同类型的别墅。所有别墅都对巴厘岛固有的别墅群风格进行了现代的诠释——每栋别墅都由两三个楼阁组成，有的采用白茅屋顶，有的采用木瓦屋顶，各个楼阁将一个庭院和一个小型健身游泳池包围其中。客厅设计结合了高高的斜屋顶和奶油色的大理石地板。室内摆放的装饰品营造出了纯净的地中海氛围：质地丰富的软垫家具都采用巧克力色、肉桂色、黄铜色、青铜色和琥珀色等相同色系的颜色，并采用当地的工艺品和古典的手工制品加以装饰。所有卧室都配备双人间的半露天浴室，浴室内摆放着富丽堂皇的独立式浴盆。大公馆位于悬崖顶端的边缘处，大公馆的地面泳池仿佛与大海直接相连。

迪马莱招牌餐厅被设计成了一个悬浮于空中的观景台，漂浮于海浪拍打的岩石上方的85米处，迪马莱餐厅是巴厘岛拥有最显赫地理位置的餐厅之一。神殿休闲吧位于迪马莱餐厅的屋顶上方，空间中弥漫着浓浓的北美氛围。奢华的彩色长沙发、水烟瓶和巨大的摩洛哥彩色玻璃灯笼打造了一个城堡般的背景。纳莫斯海滩俱乐部是一栋私人的豪华建筑，它结合了希腊群岛的浪漫气息。卡玛水疗中心耸立在海边的岩石边上，俯视大海。

1. The villas pool
2. The villas pool
3. Temple Lounge

1. 别墅的游泳池
2. 别墅的游泳池
3. 神殿休闲吧

1. Lobby
2. Car Park
3. Gymnasium
4. Karma Spa
5. Di Mare Restaurant
6. Temple Lounge
7. Veritas Wine Bar
8. Nammos Beach Club
9. Pura Matsuka Temple
10. Temple Car Park
11. Kid's Club
12. Library
13. Pool
14. Kids Pool
15. Hill Tram
16. Beach Steps
17. Indian Ocean

1. 大堂
2. 停车场
3. 健身房
4. 卡玛水疗中心
5. 迪马莱餐厅
6. 神殿休闲吧
7. 真理酒吧
8. 纳莫斯海滩俱乐部
9. 马苏卡神殿
10. 神殿停车场
11. 儿童俱乐部
12. 图书馆
13. 游泳池
14. 儿童游泳池
15. 山上电车
16. 海滨台阶
17. 印度洋

3

1. Temple Lounge
2. Resort plan
3. Restaurant and lounge

1. 神殿休闲吧
2. 度假村平面图
3. 餐厅和休闲吧

1. Di Mare Restaurant interior
2. Di Mare Restaurant exterior
3. Grand Cliff Front Residence pool area

1. 迪马莱餐厅室内
2. 迪马莱餐厅室外
3. 大悬崖别墅的泳池区域

1. Grand Cliff Front Residence exterior
2. Grand Cliff Front Residence bedroom
3. Grand Cliff Front Residence bedroom
4. Grand Cliff Front Residence living room

1. 大悬崖别墅外部
2. 大悬崖别墅卧室
3. 大悬崖别墅卧室
4. 大悬崖别墅客厅

4

1. The villa's exterior
2. The villa's living room
3. The villa's living and dining room
4. The villa's bedroom

1. 别墅外部
2. 别墅客厅
3. 别墅客厅和餐厅
4. 别墅卧室

AYANA Resort and Spa Bali

巴厘阿雅娜水疗度假酒店

Rebrand date: 2009
Location: Jimbaran Bay, Bali, Indonesia
Designer: Wimberly Allison Tong & Goo
Photographer: AYANA Resort and Spa Bali
Area: 770,000 sqm

更名时间：2009年
项目地点：印度尼西亚，巴厘岛，金巴兰湾
设计师：Wimberly Allison Tong & Goo建筑师事务所
摄影师：巴厘阿雅娜水疗度假酒店
项目面积：770,000平方米

Named after a "place of refuge" in Sanskrit, AYANA is perched on limestone cliffs up to 30 metres above the Indian Ocean near Jimbaran Bay on Bali's south-western peninsula. The property enjoys majestic views and a secluded location across its 1.3-kilometre coastline.

Designed to ensure maximum seclusion for every guest, these 78 free-standing, cliff-top Bali luxury villas are set in traditional Balinese compounds with private pools surrounded by tropical gardens. Located along the cliffs just south of the main hotel, AYANA's 38 one-bedroom Cliff Villas represent a "resort within a resort". These exquisite Bali luxury villas enjoy their own dedicated lobby surrounded by a lotus pond, which overlooks a two-tiered, infinity-edged freshwater pool and library/

recreation lounge.

Situated on 3,000 square metres of landscaped tropical gardens on the edge of the cliff, the three-bedroom Bali Luxury Villa is true to its name as a "place of refuge". Offering the most stunning cliff-top views on the island and the privacy of its secluded location in a tranquil corner of the property, this is AYANA Resort and Spa's most lavish venue for an exclusive escape. A private driveway provides exclusive and secure access, leading to a terraced water feature reminiscent of Bali's famed rice fields, with fountains that gently flow towards carved doors offering entry into the stunning villa. Inside, Balinese architecture and décor harmoniously blend with the most innovative modern comforts, top-of-the-line amenities and state-of-the-art electronics.

Resort highlights include the 22,000-square-metre Thermes Marins Thalasso Spa with one of the world's largest Aquatonic Seawater Therapy Pools and the Spa on the Rocks Villas set amidst the Indian Ocean. There are five purpose-built wedding venues: the avant-garde Champa Garden for grand-scale events, four freshwater swimming pools including a children's pool; a salt-water, infinity-edged Ocean Beach Pool at the cliff's base, accessed via an inclinator; 13 dining venues; 18-hole golf putting course; tennis pavilion; jogging/cycling path; fitness centre; children's centre; boutiques; comprehensive meeting facilities and business centre; and resort-wide WIFI.

1. AYANA Resort aerial view
2. View from AYANA lobby

1. 阿雅娜度假村鸟瞰图
2. 阿雅娜大堂外景色

巴厘阿雅娜水疗度假酒店的名字来自梵语中的"庇护所"。该度假酒店地处印度洋海平面上方30米处的石灰石悬崖，在巴厘岛西南半岛的金巴兰湾附近，横跨1.3千米的海岸线，周围景色迷人，隐蔽僻静。

为保证每位游客能享受到最佳的私人时光，78个位于悬崖顶端的、独立的豪华别墅采用巴厘岛别墅群的传统风格建造，每栋别墅都带有一个私人游泳池，周围热带园林环绕。巴厘阿雅娜水疗度假酒店的38个一居室悬崖别墅位于酒店主建筑南端的悬崖上，是"度假村中的度假村"。这些精致豪华的巴厘岛别墅拥有其雅致的大堂，大堂外围是一个莲花池，莲花池俯瞰双层的无边际净水池和图书馆（或休闲酒吧）。

三居室巴厘岛豪华别墅坐落于悬崖边上3,000平方米的热带景观园林中，是一个名副其实的"庇护所"。这里拥有岛上最迷人的绝壁景观和度假村中最隐蔽的地理位置，极大地保证了游客的隐私，这里是巴厘阿雅娜水疗度假酒店为私人旅行打造的最奢华的场馆。游客可沿私人车道，安全隐蔽地进入，然后到达一个梯田状的水景，水景仿造巴厘岛著名的稻田建造，水景中的喷泉喷出的水形成涓涓细流，流向雕刻门，这些雕刻门便是迷人的别墅入口。度假村中，巴厘岛风格的建筑和装饰与创新型的现代舒适理念、国际顶尖的设施和先进的电器设备巧妙融合。

巴厘阿雅娜水疗度假酒店最突出的设计还包括22,000平方米的Thermes Marins Thalasso水疗中心，该中心拥有世界上最大的海水疗养池和水疗中心，疗养池和水疗中心都位于印度洋上的岩石别墅中。

此外，还有5个专门建造的婚宴场地，其中时尚的占城花园专为盛大聚会精心打造；包括一个儿童游泳池在内，共有四个淡水游泳池；游客可通过观光电梯到达崖壁上方的无边际海洋泳池；还有13个进餐场馆、18球洞的高尔夫球场、网球场、适合慢跑或自行车运动的小路、一个健身中心、一个儿童活动中心、多个精品店、综合型会议设施、商务中心和遍布整个度假村的无线网络。

1. Langit Theatre
2. Sami Sami
3. Padi Restaurant
4. Damar terrace
5. Kisik seafood bar & grill
6. Honzen Japanese Restaurant
7. Yoga pavilion
8. Spa café
9. H2O
10. The Martini club
11. Dava Restaurant
12. C-bar
13. The Rock Bar
14. Upper level pool
15. Lower level pool
16. Children's pool & waterslides
17. Billiards & Table tennis
18. Kids club
19. Tennis courts
20. Fitness centre
21. Spa & fitness activity desk
22. Golf putting course
23. Spa reception
24. Aquatonic seawater therapy pool
25. River pool
26. Spa on the Rocks
27. Kubu beach
28. Ocean beach pool
29. AYANA wedding gazebo

30. Tresna bridal preparation villa
31. Tresna wedding pavilion
32. Bale Kencana
33. Astina bridal preparation villa
34. Astina wedding pavilion
35. Asmara
36. Asmara bridal suite
37. White Door wedding boutique
38. Shopping arcade
39. Resort boutique
40. Spa boutique
41. Cliff boutique & arcade
42. Jimbaran lawn
43. Couryard
44. Ballroom & function rooms
45. Business centre
46. Group arrival area
47. Resort lobby
48. Angin Aris
49. Chandra Surya
50. Bale Janji
51. Jetty
52. Alamanda room
53. Library/Business centre
54. Villa Boardroom
55. Villa lobby
56. Champa garden
57. Inclinator

1. Langit Theatre餐厅
2. Sami Sami餐厅
3. 帕迪餐厅
4. Damar Terrace餐厅酒吧
5. Kisik酒吧和烧烤
6. 本膳日本料理餐厅
7. 瑜伽馆
8. 水疗餐厅
9. H2O池畔咖啡厅
10. 马提尼俱乐部
11. Dava餐厅
12. C吧
13. 岩石吧
14. 无边泳池上层
15. 无边泳池下层
16. 儿童泳池与水滑梯
17. 桌球区
18. 儿童俱乐部
19. 网球场
20. 健身中心
21. 水疗与健身中心活动区
22. 高尔夫球场
23. Spa接待处
24. Aquatonic®海水疗养泳池
25. 悠闲河流式泳池
26. 岩石Spa
27. Kubu海滩
28. 海洋沙滩泳池
29. AYANA婚礼露台

30. Tresna新娘别墅
31. Tresna婚礼亭阁
32. Bale Kencana崖边花园
33. Astina新娘别墅
34. 新娘别墅婚礼亭阁
35. Asmara仪式亭阁
36. Asmara新娘套房
37. White Door婚礼精品店
38. 购物拱廊
39. 度假村精品店
40. Spa精品店
41. 崖边精品店
42. 金巴兰草坪
43. 中庭
44. 宴会厅与功能厅
45. 会议功能厅
46. 团体接待区
47. 度假村大堂
48. 户外草坪区
49. Chandra Surya户外会议设施
50. 宾客休息、观景区
51. 廊桥码头
52. Alamanda房
53. 图书馆 / 商务中心
54. 别墅会议室
55. 别墅大堂
56. 占巴花园
57. 斜坡

1. Resort plan
2. Ocean beach pool
3. Children's pool

1. 度假村平面图
2. 滨海泳池
3. 儿童泳池

1. Cliff Villa pool
2. AYANA Villa's pool area
3. River pool

1. 悬崖别墅泳池
2. 阿雅娜别墅泳池区
3. 悠闲河流式泳池

3

1. AYANA villa
2. Villa's lobby
3. Padi Restaurant
4. Kisik Restaurant

1. 阿雅娜别墅
2. 别墅大堂
3. 帕迪餐厅
4. Kisik餐厅

1. Lobby entrance
2. AYANA Villa living room
3. Club Room
4. Cliff Villa bedroom

1. 大堂入口
2. 阿雅娜别墅起居室
3. 俱乐部客房
4. 海景悬崖别墅卧室

Mai Khao Dream Villa Resort & Spa

麦卡奥愿景别墅温泉度假村

Completion date: 2011
Location: Phuket, Thailand
Designer: dwp
Photographer: Mai Khao Dream Villa Resort & Spa
Area: 14,400 sqm

竣工时间：2011年
项目地点：泰国，普吉岛
设计师：dwp
摄影师：麦卡奥愿景别墅温泉度假村
项目面积：14,400平方米

The words "Mai Khao", in Thai, mean "white wood", which was the defining accent for the design concept of Mai Khao Dream Villa Resort & Spa, nestled on the Northwest coast of the island. The client's vision for the resort, interpreted by world-class architecture and interior design firm dwp, was for contemporary Thai style luxury villas, aimed at a six-star experience that would emphasise prestige and elegance. The design brief included the creation of 22 exclusive opulent villas, with pools and jacuzzis, an all-day dining restaurant, clubhouse with fitness centre and a spa.

Ensuring that the resort attained an authentic Thai touch,

the brief was translated to an exquisite ethereal feel, where white is a dominant feature throughout the resort. Inspiration was primarily drawn from Thai fabrics, textures, artworks and historical drawings, as well as ocean-washed beach houses, to tie in with its seaside location. White-washed woods and natural teak met with textured cement and hints of turquoise in furniture fabrics and gold accents. When combined, all these aspects evoked luxurious Thai beach villas, with a touch of upscale New England ocean-front extravagance to their interiors.

The client's love for Thai arts and crafts led dwp to take traditional paintings from Thai myths and legends, surrounding the theme of water, and reintroduce them with a modern twist. Scenes were added as patterns on glass in the bathrooms. Arched walls and ceilings were ornately clad in the restaurant and L-shaped decorative mural panels were installed in public areas, stretching up a wall and across the ceiling. This was all achieved with a degree of added simplicity, in order not to overpower any given space.

Traditional Thai styled fixtures and fittings provided authenticity through and through, including brass lighting equipment and teak wood furniture. Iconic Thai paintings

1. Dream Bar
2. Relaxation area on the deck

1. 梦想吧
2. 平台休息区

were also featured throughout the resort. Artworks chosen for living rooms were selected to reflect the everyday life of ancient Thais, while bedroom pieces depicted more intimate settings of historical characters. Old-styled Thai art relics were placed and presented in the lobby and public areas, to add the finishing touches for guests and staff to admire and enjoy, while the spa radiates a sense of serene calm, ultimate comfort and sheer tranquility.

This 2- and 3-bedroom villa resort exudes unstinting grandeur in a tranquil and natural environment, blending perfectly with its location, right where the warm turquoise sea provocatively caresses the golden sand of Maikhao Beach.

"麦卡奥"在泰语中意为"白色的木头"，正体现了坐落在普吉岛西北海岸的麦卡奥愿景别墅温泉度假村的设计精髓。委托人授权世界级建筑和室内设计公司dwp来打造一个泰式奢华度假村，旨在为游客提供六星级的优雅和尊贵体验。设计要求打造22座独立奢华别墅（配有私人游泳池和极可意按摩浴缸）、全天候餐厅、带有健身中心的俱乐部和一个温泉水疗中心。

为了保证度假村得到泰式文化的精髓，设计营造出精致而优雅的氛围，以白色作为度假村的主色调。设计灵感主要来自于泰国的织物、材质、艺术品、历史绘画和被大海洗涤过的海滩小屋。白色的木材和天然柚木与具有纹理的水泥、蓝绿色的家具织物和金色点缀混合在一起，共同打造了奢华的泰式海滩别墅，同时又在室内糅合了新英格兰顶级的奢华海洋风。

委托人热爱泰国艺术和手工艺，因此，dwp在设计中引入了以水为主题的关于泰国神话传说的传统绘画，并在其中加入了现代色彩。浴室的玻璃上添加了图案。餐厅的拱形墙壁和天花板上装饰着L形装饰壁板，从墙面一直伸展到天花板上。于此同时，室内设计的简洁感让所有空间都不过分突出。

传统泰式装饰无所不在，黄铜灯具和柚木家具尽显泰国风情。度假村遍布典型的泰国绘画作品。起居室所采用的艺术品反映了古代泰国人的日常生活，而卧室里的艺术品则模仿了历史人物的家居背景。旧式泰国艺术品被摆放在大堂和公共区域，让宾客和员工能够尽情欣赏。温泉中心传递出宁静祥和之感，舒适而安宁。

两室和三室的度假别墅在宁静的自然环境中显得大气而宏伟。别墅四周，温暖的蓝绿色海水轻抚着麦卡奥海滩的金沙。

1. Lobby
2. Ground floor plan
3. Lobby
4. Pool area

1. 大堂
2. 一层平面图
3. 大堂
4. 游泳池

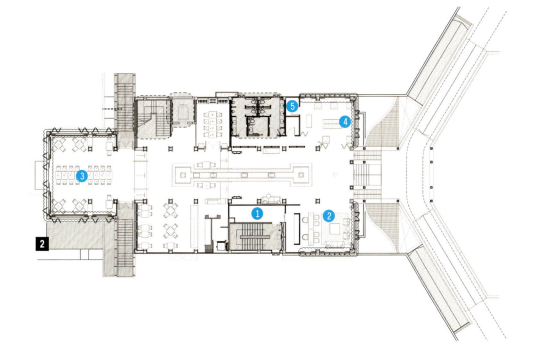

1. Front office

2. Reception

3. French restaurant

4. Retail

5. Storage

1. 前台办公室

2. 前台

3. 法国餐厅

4. 零售店

5. 储藏室

1. Clubhouse
2. First floor plan
3. Two-Bedroom Beachfront Pool Villa living and dining area
4. Two-Bedroom Beachfront Pool Villa dining area
5. Two-Bedroom Beachfront Pool Villa living area

1. 俱乐部
2. 二层平面图
3. 双卧海滨泳池别墅的客厅和餐厅
4. 双卧海滨泳池别墅的餐厅
5. 双卧海滨泳池别墅的客厅

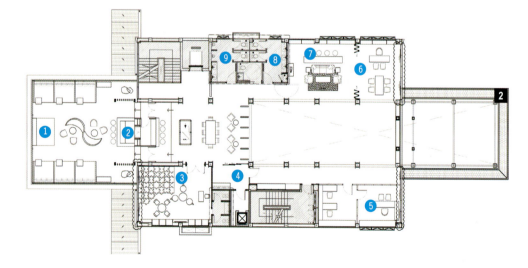

1. Stage

2. Bar

3. Kid's club

4. Pantry

5. Office

6. Business centre

7. Library

8. Female room

9. Male room

1. 舞台

2. 酒吧

3. 儿童俱乐部

4. 备餐室

5. 办公室

6. 商务中心

7. 图书室

8. 女洗手间

9. 男洗手间

1. Spa reception
2. Spa relaxation
3. Spa treatment room
4. Second floor plan

1. 水疗中心前台
2. 水疗休闲区
3. 水疗理疗室
4. 三层平面图

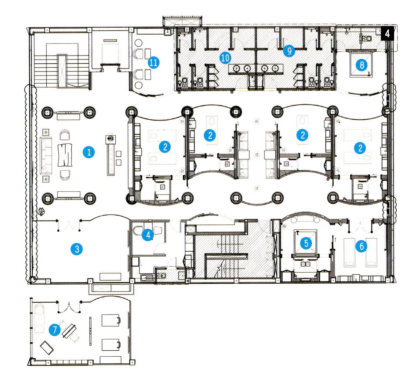

1. Reception	1. 前台
2. Treatment room	2. 理疗室
3. Fitness	3. 健身室
4. Spa manager	4. 水疗经理室
5. Jacuzzi	5. 极可意按摩浴缸
6. Spa suite	6. 水疗套房
7. Salon	7. 沙龙
8. Chromotherapy	8. 色彩疗法
9. Female locker room	9. 女更衣室
10. Male locker room	10. 男更衣室
11. Foot massage	11. 足疗

1. Villa living area
2. Villa plan
3. Villa guest bedroom
4. Villa guest bedroom

1. 别墅起居区
2. 别墅平面图
3. 别墅客卧
4. 别墅客卧

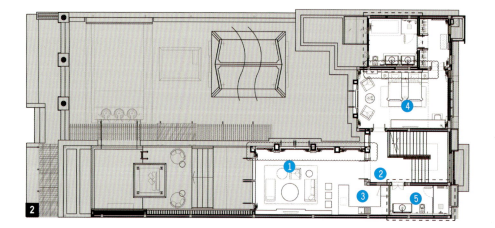

1. Living room
2. Foyer
3. Pantry
4. Bedroom
5. Bathroom

1. 客厅
2. 门厅
3. 备餐室
4. 卧室
5. 浴室

Casa de La Flora Resort

芙罗拉别墅度假村

Completion date: 2011
Location: Phangnga, Thailand
Designer: VaSLab Architecture
Photographer: Jason Michael Lang
Area: 6,500 sqm

竣工时间：2011年
项目地点：泰国，攀牙
设计师：VaS建筑事务所
摄影师：杰森·迈克尔·朗
项目面积：6,500平方米

The latest member of Design Hotel in Thailand, Casa de La Flora, the 36 cube-shaped villas located in Khao Lak, brings a modern edge to this palm tree beach of Phangnga province. Designed by VaSLab Architecture, commissioned in 2008 by one of leading Thai businessmen Sompong Dowpiset, this beachfront resort was aimed to serve as a new high profile but yet humble destination hotel in this beautiful town of the southern Thailand.

The brief given to the architect is a unique resort that consists a series of pool villas with maximum ocean views possible. Facilities as reception lounge, swimming

pool, pool bar, beachfront restaurant, spa, fitness, and library are the must-have programmes in this hotel. The owner challenges the architect to create a bold look of architecture but yet yields to warmness and nature after its implied name, "flora". VaSLab's metaphorical design takes on the act of "arising flora", where each concrctc versus wood villa reflects as a flora form, emerges from the ground, and blooms to reach the daylight. Deviated walls and tilted roofs are characterised throughout the series of 36 cubic-form villas, where these tapered elements do not only recall the act of arising flora but they widen the rooms' perspective frames when looking outward to the sea.

The continuity of these lines can be seen also in interior space and at interior elements such as built-in beds, coffee tables, and built-in cabinets. Custom-made furniture designed by Anon Pairot Design Studio carries this thematic design as some of them represent organic form of a flora. The same as landscape and hardscape work from a talented designer T.R.O.P., who extends the lines of architecture into a set of charming path ways, pavement blocks, green walls, etc., as they act as its architecture's root, stem, and branches. APLD, the lighting designer, abstractly sets the resort's lighting

1. View from the ocean
2. Deviated roof lines, metaphorical design to reflect "the arising flora"
3. Sun deck and wooden screen wall at beachfront villas

1. 从海上看度假村
2. 偏离的屋顶线、富有隐喻的设计都反映了 "破土的植物" 这一理念
3. 海滨别墅的日光浴平台和木制屏风

to provoke the main architectural elements: deviated walls, tilted roofs, as if the villa cubes are arising above the ground.

Its glass-fronted villas have clean interiors featuring concrete surfaces, natural stone walls, and wooden floors/ceilings. Ten units stand directly to the beach, with maximum sea views, and all come with private pools, 24-hour butler service and the latest in-room entertainment. Eco-friendly credentials come in the form of an ozone (low-chemical) purification system for the swimming pools and waste-water and rain water recycling.

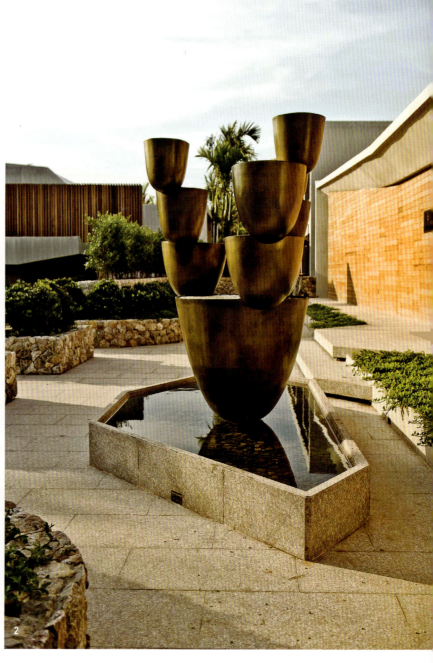

作为泰国设计酒店的最新成员，芙罗拉别墅度假村的36座方形别墅坐落在窊立，为这个攀牙省的棕榈沙滩注入了现代气息。2008年，泰国著名商人松庞·道皮斯特委托VaS建筑事务所对这个滨海度假村进行设计，目的是在这个美丽的泰国南部小镇打造高端而低调的度假酒店。

设计师被要求打造一个独特的度假村，里面的泳池别墅拥有最大化的海洋美景。前台休息室、游泳池、泳池吧、海滨餐厅、水疗馆、健身房和图书馆等设施都是酒店内的必备项目。酒店所有人要求建筑师为建筑打造一个大胆的外观，同时又要体现酒店名称"芙罗拉"（英文原意为"植物"）的温暖与自然。VaS事务所的隐喻设计以"破土的植物"为概念，每座混凝土和木材建造的别墅都反映了一种植物从地面破土而出，并且在阳光中绽放。偏离的墙壁和斜屋顶是36座别墅的统一特色。这些锥形元素不仅反映了"破土的植物"这一概念，还拓宽了房间的视野框架，使其能够更好地享受海洋的风景。

这些线条在室内空间的嵌入式大床、咖啡桌和嵌入式壁橱中得到了延续。由阿农·派洛特设计工作室特别定制的家具延续了这一主题设计，一些家具展现了植物的有机造型。富有天赋的设计师T.R.O.P.为度假村进行了景观和硬景观的设计，将建筑的线条拓展到一系列迷人的小路、铺面和绿色墙壁中，让它们作为建筑物的根、茎、枝叶。APLD事务所打造的灯光设计突出了主要建筑元素：偏离的墙壁、斜屋顶，让别墅看起来仿佛破土而出一样。

正面为玻璃的别墅拥有简洁的室内设计，以混凝土表面、天然石墙和木制地板/天花板为特色。十座套房直面海滩，享有最大限度的海景，全部设有私人游泳池、24小时管家服务和最新的房间娱乐设施。游泳池的臭氧净化系统以及废水、雨水收集系统为建筑带来了环保价值。

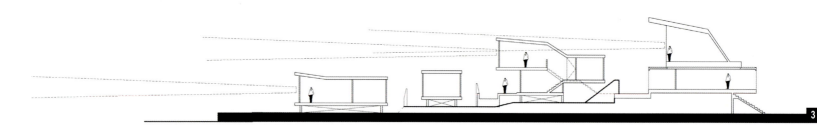

1. Behind a private gate at garden-view villa
2. A sculpture with landscape in front of spa building
3. Section
4. Beachfront villa with inside-out relationship between interior and exterior
5. Beachfront Grand Pool Villa plan

1. 私人大门后方的花园别墅
2. 水疗中心前方的雕塑和景观
3. 剖面图
4. 海滨别墅的室内外空间融合了起来
5. 海滨高级泳池别墅平面图

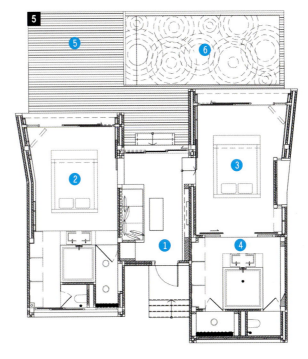

1. Living room
2. Bedroom
3. Master bedroom
4. Bathroom
5. Decks
6. Pool

1. 客厅
2. 卧室
3. 主卧
4. 浴室
5. 平台
6. 游泳池

1. Reception
2. Spa room
3. View from entry level
4. Studio Pool Villa plan

1. 前台
2. 水疗室
3. 入口层
4. 泳池别墅平面图

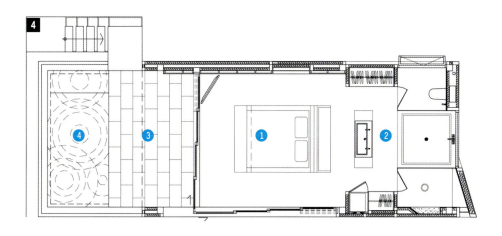

1. Bedroom
2. Bathroom
3. Terrace
4. Pool

1. 卧室
2. 浴室
3. 露台
4. 游泳池

1. Beachfront living room
2. Duplex villa living room
3. Master bedroom
4. View of master bathroom

1. 海滨别墅客厅
2. 豪华别墅客厅
3. 主卧室
4. 主卧室

Hotel de la Paix Cha Am Beach, Hua Hin

七岩酒店

Completion date: 2008
Location: Cha-Am, Thailand
Designer: Duangrit Bunnag Architect Limited – DBALP
Photographer: Hotel de la Paix Cha Am Beach, Hua Hin
Area: 30,400 sqm

竣工时间：2008年
项目地点：泰国，七岩
设计师：DBALP建筑公司
摄影师：七岩酒店
项目面积：30,400平方米

The hotel concept is essentially about nature. Not about cultural locality, but purely about its natural context. The project's vocabulary requires architectural sublimity so the purity of nature can be fully appreciated.

The project is laid out in a linear approach. The sequence of space poetry begins with a geometrical, landscaped courtyard, a large white marble piazza, which brings you to grand marble steps that lead to a large open-air lobby with a latticed-wood structure spanning 18 metres; it finishes at a seemingly never-ending stretch of the reflective pool that meets the horizon. A beach front restaurant marks the end.

The function of the project is simple and straightforward: a restaurant with a beachfront experience, a spa for ultimate relaxation, a room with a view and a house with a pool. A key highlight at Hotel de la Paix Cha Am Beach is the Red Bar. Thus, from now on, you may have to consider red as a function.

The design of the project is quite a distance from the word "contemporary" or "style", if not further than the word "resort" itself. The notion of the design is about a place, its emotional and spatial connotations.

How does the décor pay homage to the Thai culture? The notion of Thai culture can lead to a diverse and contradictory conversation. What is the actual definition of "Thai culture"? Is it something you see or something you taste and feel? This project can perhaps embody Thai culture through the analogy of Thai food: it is not about how it looks, but more about how it feels.

The architects designed most of the furniture themselves because the strong personality of the architecture suits few furniture collections. All the material used for furniture and finishing are inspired by nature; thus, they have stone, wood, glass, concrete and red. Red can also be considered as a material too.

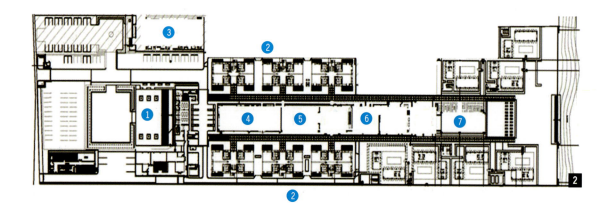

1. Lobby	1. 大堂
2. Building	2. 建筑
3. Event Centre	3. 活动中心
4. Spa Indochine	4. 印度支那水疗中心
5. Chill pool	5. 冷水泳池
6. Library	6. 图书室
7. Active pool	7. 活动泳池

1. Beach
2. Resort plan
3. Chill pool and Red Bar
4. Cabana dining

1. 沙滩
2. 度假村平面图
3. 冷水游泳池和红吧
4. 凉亭就餐

1. Active pool
2. Event centre exterior
3. Clouds loft

1. 活动游泳池
2. 活动中心外观
3. 云彩阁楼

七岩酒店以自然为经营理念。酒店不以当地文化为重点，而是完全依存当地的自然环境。酒店要求建筑具有高贵的气质，以凸显自然的纯粹之感。

项目呈线性展开。酒店呈几何造型的景观庭院和巨大的白色大理石广场引领着人们走向巨大的大理石台阶，前往横跨18米的木格结构露天大堂。大堂后面是一个看似永无尽头的倒影池，直到地平线。空间末端是一个海滨餐厅。

项目的功能简单而直接，包括海滨餐厅、奢华休闲水疗馆、景观客房和游泳池别墅。七岩酒店的特色是红色酒吧。因此，红色在酒店内异常显眼。

项目设计既不"现代"也不"时尚"，甚至偏离的"度假村"的典型形象。设计概念以空间情绪和空间内涵为主题。

如何让装饰向泰国文化表达敬意？泰国文化能够引出各种各样的矛盾对话。"泰国文化"的真正定义又是什么？是能看见、能感觉的吗？项目通过泰国美食来诠释泰国文化：这无关外观，而是一种内在感觉。

建筑师亲手设计了大多数家具，因为建筑强烈的个性难以与普通家具相匹配。家具和装饰所采用的材料都以自然为灵感。因此，他们运用石头、木材、玻璃、混凝土和红色作为材料。在这里，红色也是一种"材料"。

1. Pool Villa exterior
2. Pool Villa exterior
3. The Spa Indochine – Hydra Pool

1. 泳池别墅外观
2. 泳池别墅外观
3. 印度支那水疗馆——海德拉游泳池

1

2

1. Library
2. Pool Villa bedroom

1. 图书室
2. 泳池别墅卧室

1. Horizon Room
2. Pool Villa bedroom
3. Pool Villa bathroom
4. Horizon Room bathroom

1. 地平线房
2. 泳池别墅卧室
3. 泳池别墅浴室
4. 地平线房的浴室

Asara Villa & Suite

阿萨拉别墅度假村

Completion date: 2007
Location: Prachuabkirikhun, Thailand
Designer: Tanitpong Chalermpanth (Principal),
Visoot Thubthimthed (Design Director)
Photographer: Chanok Thammarakkit, Ocean Property Co., Ltd.
Area: 12,730 sqm

竣工时间：2007年
项目地点：泰国，普拉齐亚吉里昆
设计师：塔尼特朋·查勒姆潘斯（主持设计）、
韦苏特·萨西姆斯德（设计总监）
摄影师：查诺克·萨马拉奇特、海洋房产公司
项目面积：12,730平方米

Located near the world-renown seaside resort town of Hun Hin, the Asara Villa & Suite is a five-star resort comprising 2 luxurious beachfront pool villas, 50 private pool villas, and 44 suites nestled along a man-made lagoon and swimming pool, away from the hustle and bustle of the town itself.

The design accommodates a long, narrow site with minimal ocean frontage. Master planning concept creates an internal focus with man-made lagoons and strives to maximise favourable views, either of the ocean or of the internal lagoons, from all points while accommodating stringent ocean-front development regulations.

Existing natural channel at the centre of the site has been expanded and developed into focal water features, combining picturesque lagoons with swimming pools, creating a refreshing "spine" of water which appears to flow from one end of the site down to the sea. Earth from the lagoon excavation was used to elevate rear portion of site to enhance views to the sea.

Maximised retention of natural features of site along with existing vegetation is to enhance the built environment. Existing specimen raintree has been used as the primary focus in design of the beachfront restaurant. The beachfront restaurant is broken up into a collection of smaller pavilions, while the feature fine-dining restaurant is located just one step away from the beach and purposely elevated to maximise ocean views, stacked above a large podium below which houses kitchen/restaurant/back-of-house functions/conference hall, to minimise massiveness and to accommodate the natural seaside setting along with an existing specimen Gampu tree.

Main lobby/reception pavilion is elevated high above the natural ground to capture both distant views of the ocean, as well as close-up views of adjacent tropical lagoon/gardens. The main pavilion also houses a small

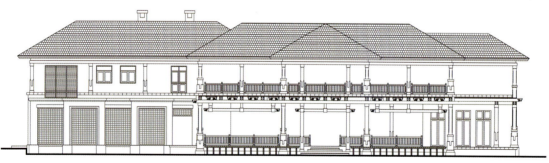

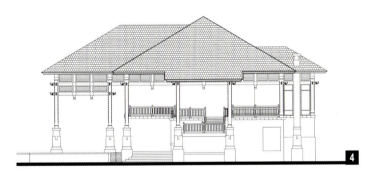

1. Beachfront swimming pool
2. All-day dining – elevations
3. Pavilion at beachfront swimming pool
4. Reception and administration – elevation

1. 海滨游泳池
2. 全日餐厅——立面图
3. 海滨游泳池畔的凉亭
4. 前台和行政区——立面图

restaurant/retreat adjacent to the tropical courtyard.

The distinctive form and character of the architecture of the resort have been inspired by the old beach cottages of Hun Hin, which were the popular vacation homes of Bangkok's aristocracy over 100 years ago and the traditional summer palace of royalty located nearby, and has been integrated with contemporary Thai design.

The distinctive multiple-pavilion villas provide a relaxing and private retreat for the guest, with large terraces offering sweeping panoramic vistas of the ocean and lagoons. Luxury villa units consist of independent living pavilions and sleeping pavilions.

Earth-tone colours, texture of cedar wood shingle roofing present a rustic appearance, allowing architecture to blend into the natural tropical environment. The cottage's roof forms, distinctively-crafted columns, expansive terraces, and decorative plaster walls combine to add a touch of flair and elegance to the exotic oriental character.

紧邻举世闻名的浑欣海滨度假胜地，五星级阿萨拉别墅度假村由两座奢华的海滨游泳池别墅、50套私人游泳池别墅和44套沿着人工湖和游泳池而建的套房组成，远离喧嚣的城市。

项目位于海边一块狭长的场地上，总体规划打造了一个向内聚焦的人工湖。在严格遵守海岸开发条例下，力求最大限度地利用海洋和人工湖的风景。

场地中央的天然海峡被扩展开发成为一个中央水景，结合了风景如画的人工湖和游泳池，营造出贯穿场地、直达海洋的清新水道。人工湖挖掘出的土被用于垫高场地后方，以更好地享有海景。

对场地自然特征的最大化保护提升了建筑环境。原有的雨树曾是海滨餐厅设计的焦点。海滨餐厅由一系列小亭子组成。特色精品餐厅则离沙滩仅几步之遥，垫高的地势

可以享受更好的海景，巨大的底座部分里设置着厨房、餐厅、后台设施和会议厅，尽量减少建筑体量，与周边的自然景色融为一体。

大堂/接待处高出自然地面许多，能够捕捉到远处海洋和近处热带人工湖/花园的景色。主亭里同样设有一家小餐厅，紧邻热带庭院。

度假村建筑独特的造型和风格受到了浑欣古老的沙滩小屋（百年前，沙滩小屋曾是曼谷贵族的度假别墅）和附近的泰国皇室夏宫的启发，同时又增添了现代泰式设计。

独特的多亭式别墅为宾客提供了轻松的私人领域，巨大的平台提供了海洋和人工湖的壮丽美景。奢华的别墅由独立的起居亭和安睡亭组成。

大地色调的雪松木瓦屋顶呈现出质朴的外观，让建筑与天然的热带环境融为一体。别墅的屋顶造型、精致的柱子、广阔的平台和装饰石膏墙共同为东方风情增添了质感。

1

1. View from Presidential Villa
2. Beachfront swimming pool
3. Resort plan

1. 从总统别墅向外看
2. 海滨游泳池
3. 度假村平面图

1. One-bedroom suite	1. 一室套房
2. One-bedroom pool villas	2. 一室泳池别墅
3. Two-bedroom pool villas	3. 两室泳池别墅
4. All-day dining restaurant	4. 全日餐厅
5. Reception	5. 前台
6. Beachfront swimming pool	6. 海滨游泳池
7. Presidential Villa	7. 总统别墅
8. Two-bedroom beach villa	8. 两室海滨别墅
9. Guest parking area	9. 客用停车场
10. Spa	10. 水疗中心

1. Lobby & reception area
2. Lobby & reception area
3. Courtyard garden
4. Reception and administration – first floor plan

1. 大堂和前台
2. 大堂和前台
3. 庭院花园
4. 前台和行政区——二层平面图

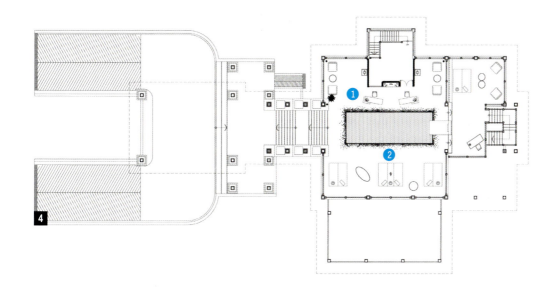

1. Reception
2. Sitting area

1. 前台
2. 休息区

1. Siam Bistro Restaurant
2. Kampu by Choice Restaurant
3. All-day dining – ground floor plan
4. Siam Bistro Restaurant
5. All-day dining – first floor plan

1. 暹罗餐厅
2. 康普自选餐厅
3. 全日餐厅——一层平面图
4. 暹罗餐厅
5. 全日餐厅——二层平面图

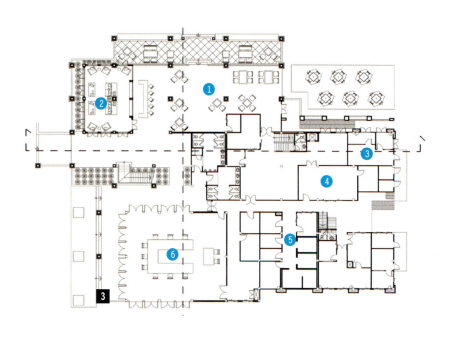

1. Specialty Bar & Grill

2. Library

3. Services

4. Kitchen

5. Back-of-House

6. Banquet/Conference

1. 特色酒吧与烧烤区

2. 图书室

3. 传菜区

4. 厨房

5. 后场区域

6. 宴会厅/会议室

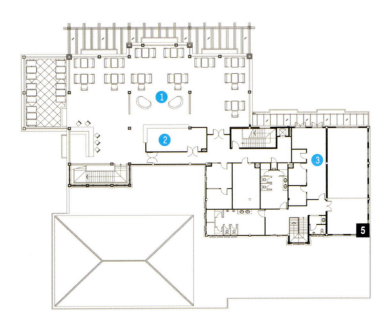

1. Kampu by Design Restaurant
2. Show kitchen
3. Back-of-House

1. 康普设计餐厅
2. 开放式厨房
3. 后场区域

1. Beachfront all-day dining restaurant
2. Entrance to Asara Spa
3. One-bedroom pool villas plan
4. One-bedroom suite plan
5. Presidential Villa living room
6. Presidential Villa bedroom

1. 海滨全日餐厅
2. 阿萨勒水疗中心入口
3. 一室泳池别墅平面图
4. 一室套房平面图
5. 总统别墅客厅
6. 总统别墅卧室

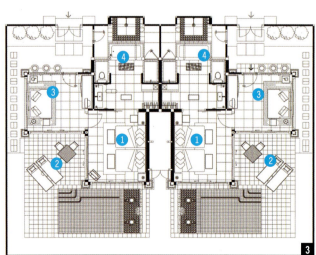

1. Bedroom   1. 卧室
2. Terrace   2. 露台
3. Living room 3. 起居室
4. Bathroom   4. 浴室

1. Bedroom   1. 卧室
2. Bathroom   2. 浴室
3. Terrace   3. 露台

Sirarun Resort

希拉隆度假村

Completion date: 2010
Location: Tubsake, Thailand
Designer: ARJ Studio Co., Ltd.
Photographer: ARJ Studio Co., Ltd.
Area: 7,600 sqm

竣工时间：2010年
项目地点：泰国，塔博塞克
设计师：ARJ工作室
摄影师：ARJ工作室
项目面积：7,600平方米

A small southern tropical landscape style resort nestled in a peaceful beach of Prachuab Keerikan province, overlooking to spectacular view of a tiny coral island, Lam la. Owned and operated by the Prombot family, the unique property accommodates 16 standard villas, 4 superior suites, private residence, along with an open-air restaurant, spa, sport centre and pools. Both sizes of accommodation are suitable for families, group of friends and couples, while its compact size ensures everyone enjoys some privacy and a real sense of being in a retreat. The resort is a bold but essential marriage of the traditional and modern. The main design concept was to

represent the blending contemporary Asian touch with the local simple characteristics of space, colour and elements. Local materials, such as bamboo, rattan, local stone, and teak wood, are carefully mixed to express their natural surface into modern artistic. To complement more earthy design, all villas come with their own enclosed private lush gardens and outdoor shower/Jacuzzi.

The overall layout sets each building far apart so as to gain more privacy and preserve the trees. The open-air lobby is located to the highest terrain overlooking the magnificent vista of Tub Sakae bay one side and the mountain on the other. The giant-size infinity-edged

swimming pool and terrace which are divided in three parts are clustered by the main lobby, restaurant and pool villas to form a self-sufficient little community. Within the compound, kids can enjoy in the sand pond or splash in the child pools while adults soak in the sun, admire the sweeping views of the white-sand beaches.

The main restaurant, seating 40, located next to a bamboo-sheltered deck and pool bar & lounge, giving a sense of relax and refinement. The main pavilion entrance is an elegant combination of a small floating gazebo and reflecting pond, expressing the perfectly complemented resort's luxurious look and feel. The interior of each

1. Main entrance
2. Lobby and front office
3. Main dining
4. Spa
5. Main kitchen
6. Office
7. Fitness
8. Service quarter
9. Reflecting pool
10. Children pool
11. Pool bar
12. Pool
13. Outdoor dining
14. Owner's House
15. Pool terrace
16. Pool Villa A
17. Pool Villa B

1. 主入口
2. 大堂和前台办公室
3. 主餐厅
4. 水疗中心
5. 主厨房
6. 办公室
7. 健身中心
8. 服务区
9. 倒影池
10. 儿童泳池
11. 泳池吧
12. 游泳池
13. 露天餐厅
14. 业主之家
15. 泳池平台
16. 泳池别墅A
17. 泳池别墅B

1. Resort pool
2. Resort plan
3. Resort garden

1. 度假村游泳池
2. 度假村平面图
3. 度假村花园

villa has a distinct visual style, with a unique colour scheme that ranges from orange to yellow, burgundy and brown. These come together in a complementary colour palette to endow the estate with the tranquil mood.

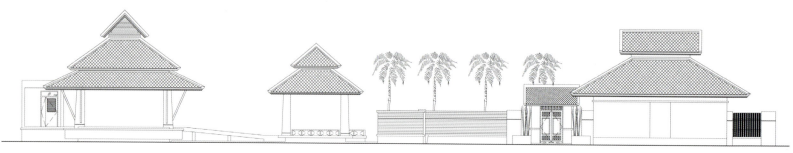

1. Exterior view of the villa
2. Overall elevations
3. Exterior view of the villa

1. 别墅外观
2. 整体立面图
3. 别墅外观

这座小型南半球热带景观度假村坐落在泰国普拉查布·吉里坎省宁静的海滩上，远眺微型珊瑚岛拉姆拉的绝美景色。度假村由普洛姆波特家族所拥有，共有16座标准别墅、4套高级套房和私人住宅，配有露天餐厅、温泉、运动中心和泳池等设施。住宿设施的规模适合家庭、友人和情侣，其紧凑的结构保证了个人隐私，提供了真正的度假感。

酒店大胆地结合了传统和现代风格。主要设计理念是呈现现代亚洲风格与当地简单的空间、色彩、元素设计的结合。竹子、藤条、石头、柚木等当地材料被精心混合在一起，在自然基础上体现了现代美学。为了完善质朴的设计，所有别墅都配有郁郁葱葱的私人花园和露天淋浴间或极可意浴缸。

每座建筑相隔甚远，以获得更多的隐私并保护树木。露天大堂位于最高处，一面俯瞰着塔博萨奇湾的壮丽景色，一面享有山脉的美景。大堂、餐厅和泳池别墅中央是三个大型无边界泳池和平台，形成了一个自给自足的小型社区。在这里，孩子们可以在沙池和儿童泳池中尽情嬉戏，而大人则能够沐浴阳光、欣赏白沙滩的美景。

主餐厅可同时容纳40人，设在竹子平台和泳池酒吧的旁边，散发出放松感和精致感。主亭入口优雅地结合了小型漂浮瞭望台和倒影池，完美地展现了度假村的奢华感。每座别墅的室内都有其独特的风格，享有独一无二的色彩主题，从橘黄、黄色，到紫红、棕色。这些颜色汇集到一起，共同组成了互补的色彩搭配，为度假村带来了宁静的氛围。

1. Detail of the villa
2. Detail of the villa
3. Villa exterior pool
4. Corridor
5. Overall plan

1. 别墅细部
2. 别墅细部
3. 别墅的露天泳池
4. 走廊
5. 整体规划图

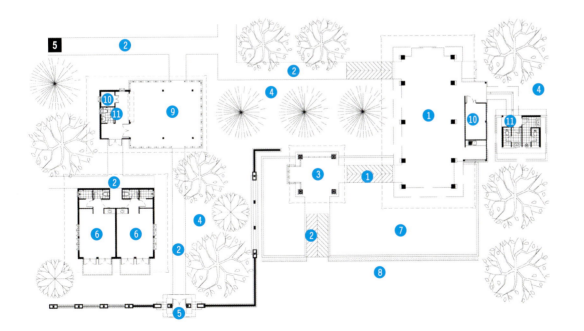

1. Main lobby	1. 主大堂
2. Walk way	2. 走道
3. Sa-la (Entry)	3. 入口
4. Garden	4. 花园
5. Sub gate	5. 侧门
6. Spa	6. 水疗中心
7. Reflecting pool	7. 倒影池
8. Parking	8. 停车场
9. Lobby spa	9. 大堂水疗
10. Storage	10. 储藏室
11. Toilet	11. 洗手间

1. Bar
2. Reception
3. Restaurant
4. Restaurant

1. 酒吧
2. 前台
3. 餐厅
4. 餐厅

1. Living room
2. Guestroom
3. One-bed guestroom
4. Two-bed guestroom
5. Floor plan

1. 客厅
2. 客房
3. 单人客房
4. 双人客房
5. 平面图

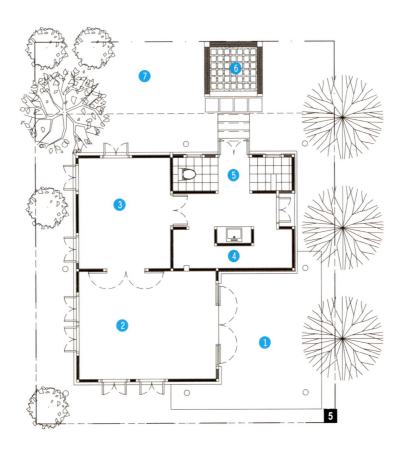

1. Terrace
2. Living room
3. Bedroom
4. Walk-in closet
5. Toilet
6. Outdoor shower
7. Garden

1. 露台
2. 起居室
3. 卧室
4. 换衣室
5. 卫生间
6. 户外淋浴
7. 花园

The Sofitel So Mauritius Bel Ombre

索菲特毛里求斯贝尔欧布莱度假村

Completion date: 2010
Location: Bel Ombre, Mauritius
Designer: Architect Lek Bunnag
Photographer: Ludovic Aubert, Lek Bunnag, Sylvie Becquet, Jacques Rocca-Ferra
Area: 130,000 sqm

竣工时间：2010年
项目地点：毛里求斯，贝尔欧布莱
设计师：莱克·巴纳格
摄影师：卢多维克·奥贝尔特、莱克·巴纳格、赛尔维·贝克特、
雅克·罗卡–费拉
项目面积：130,000平方米

Imagined by architect Lek Bunnag, these 92 suites and villas, all single-storey, are nestled in a lush natural setting. These islets bathed in light, protected in a green paradise, provide intimacy with unlimited comfort. The sleek, rounded architecture offers a naturally comfortable, fluid space. The sobre design and noble materials subtly combine with decorative elements designed by Kenzo Takada, with their colourful, floral motifs reflecting the natural environment. For an intense getaway sensation, this tropical setting includes private outdoor bathtubs (or pools) and showers.

The private gardens and patios open out onto lush

vegetation, is an invitation to contemplation. The sleek lines of the 84 Suites (60 sqm), 6 Beach Villas (100 sqm) and 2 Beaulieu Villas (200 sqm) are perfectly integrated into the environment, blurring the lines between indoors and outdoors. The view of the gardens and patios are an invitation to engage in the art of taking it easy – a unique encounter between the French "art de recevoir" and some of the best service in the world.

The prestige suites are magnificent, independent pavilions which are superbly located in a natural environment, with private garden with bathtub and shower on the patio, adjoining bathroom. Eight of them are connected (without outdoor bathtubs in the garden – bathtub in the bathroom), each with two single beds. The two bedrooms have a single access door for constant, safe connection. The beach villas are magnificently located in a natural environment, with view over the Ocean, private garden with an 8-square-metre pool, shower on the patio and hospitality products. Villa Beaulieu is superbly located in a natural environment, with view over the Ocean, sitting room, dining room, private garden with a large, 27-square-metre swimming pool equipped with a "child safety" alarm and a private hammam.

The waterside "Le Flamboyant" restaurant puts on quite

a show with its open kitchen and proposes the best of French and Mauritian cuisine in a sophisticated setting. For a lounge ambiance, go to "La Plage" restaurant. The "So Spa" has adopted the Mauritian culture and the surrounding natural bounty. Enjoy the well-being offered by nature and your newfound vital energy!

1. Beach view
2. Resort plan
3. Villa Beaulieu's exterior
4. Restaurant

1. 海景
2. 度假村平面图
3. Beaulieu别墅外观
4. 餐厅

1. Reception pavilion
2. Spa
3. Mini club, Fitness centre and pool
4. Guest pavilion
5. "Le Flamboyant" restaurant and "Takamaka" bar
6. "La Plage" bar-restaurant

1. 前台厅
2. 水疗中心
3. 迷你俱乐部、健身中心和游泳池
4. 客房
5. 凤凰树餐厅和塔卡玛卡吧
6. 海滨酒吧餐厅

由建筑师莱克·巴纳格所设计的92套套房和别墅坐落在一片郁郁葱葱的自然环境中。这些小岛沐浴在阳光里，隐蔽在绿色的天堂中，为人们提供终极的私密感和舒适感。圆润的建筑提供了自然、舒适、流畅的空间。稳重的设计和高雅的材料与装饰元素巧妙地结合在一起。这些装饰元素由高田贤三设计，多姿多彩的图案反映了自然环境的特色。为了营造强烈的逃亡感，度假村还提供了私人露天浴缸（或游泳池）和淋浴间。

私人花园和露台一直延伸到茂密的树木之中，引人沉思。84套套房（60平方米）、6座海滩别墅（100平方米）和2座比尤利别墅（200平方米）那井然有序的线条与环境完美地结合在一起，模糊了室内外的界限。花园和露台的景色让人们感到放松——法式"接受的艺术"和世界上最好的服务邂逅了彼此。

精品套房是一系列华丽的独立凉亭，堂皇地坐落在自然环境之中。套房的浴室旁是带有浴缸和淋浴的私人花园平台。其中8套套房相互连接，各设有两间卧室。出于安全的考虑，两间卧室从一个入口进入。宏伟的海滩别墅俯瞰着海洋，其私人花园里配有8米长的泳池、露台淋浴间和相应的设施。比尤利别墅同样俯瞰着海洋，设有客厅、餐厅，其私人花园中27米长的泳池配有"儿童安全"报警设备。别墅里还设有一个私人土耳其浴室。

水畔凤凰树餐厅的露天厨房将进行独特的表演，为宾客们提供精致的法式和毛里求斯美食。如果想要酒吧间的感觉，请前往海滨餐厅。温泉水疗中心融汇了毛里求斯文化和周边的自然美景，自然健康的服务将让人恢复活力。

1. Restaurant
2. Bar and lounge
3. Bar and lounge

1. 餐厅
2. 酒吧和休闲吧
3. 酒吧和休闲吧

1. Restaurant
2. Spa and relaxation
3. Beach Villa's bathroom
4. Spa and relaxation

1. 餐厅
2. 水疗中心和休息区
3. 海滨别墅浴室
4. 水疗中心和休息区

1. Guestroom
2. Guestroom
3. Living room in the villa
4. Guestroom

1. 客房
2. 客房
3. 别墅客厅
4. 客房

Sofitel Mauritius L'Imperial Resort and Spa

索菲特毛里求斯皇室温泉度假村

Completion date: 2011
Location: Flic en Flac, Mauritius
Designer: Maurice Giraud, Artema Design (Philippe Rambaud)
Photographer: Sofitel Mauritius L'Imperial Resort and Spa
Area: 90,000 sqm

竣工时间：2011年
项目地点：毛里求斯，弗里克恩弗莱克
设计师：莫里斯·吉洛、阿尔特玛设计（菲利
普·兰巴德）
摄影师：索菲特毛里求斯皇室温泉度假村
项目面积：90,000平方米

Set amid 9 hectares of tropical gardens along a gorgeous beach of soft white sand, Sofitel Mauritius L'Imperial Resort and Spa is wholly dedicated to your comfort and pleasure. Asian-inspired architecture and sleek modern furnishings highlight the beauty of the natural setting and reveal particular attention to detail. Discerning guests will be enchanted with the graceful ambiance of the luxurious guest quarters, subtly decorated with stone and wood accents. Sofitel Mauritius L'Imperial Resort and Spa also offers wonderful restaurants in which to savour the very best of Eastern, Western, and oceanside gastronomy. Choose from among a plethora of leisure activities or simply relax and let time flow by... Whatever your

fancy, your stay is bound to be coloured with unmitigated pleasure, rich discovery, and utter well-being.

143 luxurious guestrooms and 48 suites, all featuring a balcony, private terrace, or magnificent ocean view and offering utmost comfort. Intimate havens with four-poster beds, fabric wall covering, fine woodwork, and delicate etchings for a soothing atmosphere of elegance and tranquility. Families will appreciate specially conceived suites surrounding the lovely Japanese garden. For an exceptionally grand experience, opt for a sumptuous, ocean-view Prestige Suite or the truly superlative 125-square-metre Imperial Suite. Outstandingly luxurious accommodations also feature rain showers, exclusive

Sofitel MyBeds, and personalised service around the clock.

The resort offers a fabulous choice of dining venues and bars. The Sofitel So Spa is a haven of peace, well-being, and beauty. The guests could experience absolute comfort and refinement in an Asian-inspired sanctuary. There are Hammam, beauty centre, and an audio pool, where musical vibrations enhance the relaxation experience. An exquisite beach, an overflowing pool, a fitness room, and a broad range of land and water sports will surely please you.

1. Aerial view of the resort
2. Beach view

1. 度假村鸟瞰图
2. 海滩景色

1. Beachfront pool
2. Pool and deck
3. Fountain entrance pool

1. 海滨游泳池
2. 泳池和平台
3. 喷泉水池

在9公顷的热带花园的环绕下，沿着迷人而柔软的白沙沙滩，索菲特毛里求斯皇室温泉度假村完全专注于为宾客提供舒适和愉悦的服务。亚洲风格建筑和时尚的现代装饰凸显了自然景观的美感，特别注重细节。宾客们将为以石木装点的奢华客房中优雅的气氛而着迷。索菲特毛里求斯皇室温泉度假村提供美妙的餐饮环境，让宾客尽享东西方和海滨美食。丰富的娱乐活动和简单的放松让时光飞逝。无论你抱有怎样的幻想，在度假村的生活将多姿多彩，充满了愉悦、新发现和无比的舒适健康。

143套奢华的客房和48套套房都设有阳台、私人露台，或者享有壮美的海景，为宾客提供极致的舒适之旅。这些私密的天堂配有四柱床、织物墙面、精致的木工活和精美的雕刻，形成了优雅而宁静的舒缓气氛。家庭游客将享用带有可爱的日式花园的套房。奢华的海景名品套房和125平方米的皇室套房将为宾客提供更高级的华丽体验。奢侈的套房同时还配有雨淋花洒、索菲特专属双人床和全天候专属服务。

度假村提供种类丰富的餐厅和酒吧。索菲特温泉水疗中心是平静、安康和美好的天堂。宾客将在亚洲风情的胜地中体验终极的舒适和精致。水疗中心设有土耳其浴室、美容中心和音乐泳池——音乐的共鸣将提升放松体验。精美的沙滩、充盈的泳池、健身房和一系列陆上、水上运动无疑将让人愉悦。

1. Restaurant pavilion   1. 餐厅
2. Resort plan   2. 度假村平面图
3. Banquet hall   3. 宴会厅
4. Imperial Conference   4. 皇室会议室

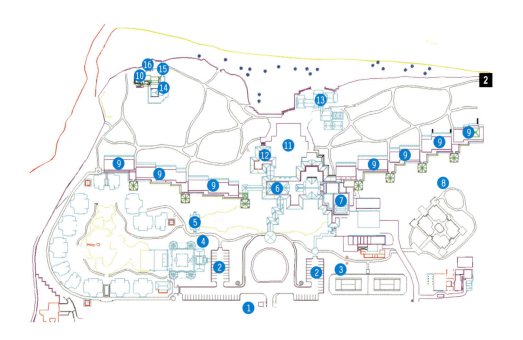

1. Main gate   1. 大门

2. Parking   2. 停车场

3. Tennis court   3. 网球场

4. Spa   4. 水疗中心

5. Gymnasium   5. 健身房

6. Lobby   6. 大堂

7. Conference room   7. 会议室

8. French garden   8. 法式花园

9. Suites   9. 套房

10. Jacaranda Restaurants   10. 蓝花楹餐厅

11. Swimming pool   11. 游泳池

12. Pink Café   12. 粉红咖啡厅

13. Tamassa bar & restaurant   13. 塔玛莎酒吧和餐厅

14. Mini club and kids swimming pool   14. 迷你俱乐部和儿童游泳池

15. Boat house   15. 游船别墅

16. Diving centre   16. 潜水中心

3

4

1. So Spa
2. Children's room
3. Prestige Suite living room
4. Family Suite parents room

1. So水疗馆
2. 儿童房
3. 高级套房起居室
4. 家庭套房父母卧室

1. Imperial Suite living room
2. Imperial Suite bathroom
3. Superior room
4. Luxury room

1. 皇室套房客厅
2. 皇室套房浴室
3. 顶级套房
4. 奢华套房

Shanti Maurice Hotel

香提莫里斯酒店

Completion date: 2010
Location: Mauritius
Designer: Jean Marc Eynaud Architects
Photographer: Jean Marc Eynaud Architects
Area: 153,800 sqm

竣工时间：2010年
项目地点：毛里求斯
设计师：珍·马克·伊诺德建筑事务所
摄影师：珍·马克·伊诺德建筑事务所
项目面积：153,800平方米

Shanti Maurice is a boutique lifestyle resort located around a pristine horseshoe coral sand cove on the largely untouched south coast of Mauritius. Set between the turquoise water of the Indian Ocean and the vivid greens of the sugarcane fields, Shanti Maurice provides through its architecture a strong tropical atmosphere conveying the sights, sounds and flavours of the unique mix of Indian, African, French and Chinese influences that form the Mauritian culture.

The hotel comprises sixty-one spacious suites and villas spread across a 15.38-hectare landscape of fragrant tropical gardens. All suites and villas are located on the beachfront with the villas offering private pools, outdoor

rain showers and dining pavilions.

Interior spaces are fluid and spacious with some large openings allowing the sunshine to bathe interiors with light. The Junior Suites are housed in villa buildings with four suites per villa. The Presidential Villa, half hidden amongst lush greenery, comprises two bedrooms with bathrooms, open courtyards and a wrap around balcony with exotically landscaped private gardens. A separate dining and living area located in front of a large swimming pool complements the unique experience of this villa.

The Spa, one of the largest in the Indian Ocean, is built around an enchanting tea pavillon surrounded by lily ponds and indigenous flower gardens. The sound of water helps a lot to create a relaxing atmosphere.

Landscaping and water form an integral part of the architecture in general, and it is particularly well represented with this project. The designers constantly tried to combine water features and interpenetration of garden into building spaces to clearly set the whole as a continuum instead of the juxtaposition of these three separate elements. In some places you cannot say if it is the garden that penetrates into the buildings or if it is the buildings that penetrate into the garden.

1. Lobby and reception
2. Shop
3. Library
4. Conference room
5. Main pool with Jacuzzi
6. Beach hut
7. Red Ginger & Lounge Bar
8. Pebbles
9. Stars
10. Fish N Rum Snack
11. Spa reception
12. Spa pool & Watsu pool
13. Yoga pavilion
14. Kid's club
15. Tennis Courts
16. Gym
17. Jogging track
18. Herb Garden
19. Jetty + Wedding pavilion
20. Car park

1. 大堂和前台
2. 商店
3. 图书室
4. 会议室
5. 带有按摩浴缸的主游泳池
6. 海滨小屋
7. 红姜休闲吧
8. Pebbles餐厅
9. 星辰餐厅
10. 垂钓餐饮区
11. Spa接待处
12. Spa泳池和水中按摩池
13. 瑜伽亭
14. 儿童俱乐部
15. 网球场
16. 健身房
17. 慢跑跑道
18. 香草园
19. 接待与婚庆亭
20. 停车场

1. Resort plan
2. Two-bedroom villa – pool terrace
3. Enchanting view

1. 度假村平面图
2. 两卧室别墅——泳池平台
3. 迷人的景色

1

1. Stars restaurant and pool
2. Stars restaurant and pool
3. Pool deck detail

1. 星辰餐厅和游泳池
2. 星辰餐厅和游泳池
3. 泳池平台细部

香提莫里斯是一个精品生活度假村，环绕着毛里求斯南海岸未经开发的珊瑚沙湾而建。香提莫里斯坐落于印度洋蓝绿色的海水和翠绿的甘蔗田之间，其建筑具有强烈的热带气息，汇集了印度、非洲、法国和中国的声、色、味，体现了独特的毛里求斯文化。

酒店共有61套宽敞的套房和别墅，分别坐落于占地15.38公顷的热带景观花园之中。所有套房和别墅都设在海岸上，别墅还配有私人游泳池、露天淋浴和就餐亭。

室内空间流畅而宽敞，巨大的窗户让阳光洒满室内。每座别墅内设有四套小型套房。在绿树掩映中的总统别墅设有两卧两卫、开敞的庭院、环形阳台和极具异域风情的私人花园。独立的餐饮起居区设在大型游泳池前方，完善了别墅的整体效果。

香提莫里斯拥有全印度洋最大的温泉馆，环绕着迷人的茶道亭，四周是莲花池和本土花园。潺潺的水声营造出更加放松的氛围。

景观和水两种元素是构成建筑的有机部分，在这个项目中的体现尤为突出。设计师一直努力将水景、花园融入建筑空间，使其成为统一的整体，而不是将它们独立起来。一些空间说不清到底是花园融入了建筑还是建筑融入了花园。

1. Pool garden view
2. Pool garden view
3. Pool garden view

1. 泳池花园
2. 泳池花园
3. 泳池花园

1. Night view of the villas
2. Night view of the restaurant

1. 别墅夜景
2. 餐厅夜景

2

1. Entrance to the lobby
2. The spacious main entrance
3. Another view of the entrance lobby

1. 大堂入口
2. 宽敞的主入口
3. 从另一个角度看入口大堂

1. Ocean views and expansive proportions of the Presidential Villa
2. Charming colours and aesthtics
3. Presidential Villa living room

1. 海洋景色和宽敞的总统别墅
2. 迷人的色彩和美学搭配
3. 总统别墅客厅

1. Guestroom   1. 客房
2. Guestroom   2. 客房
3. Guestroom   3. 客房
4. Bathroom   4. 浴室
5. Bathroom   5. 浴室

Heritage Awali Golf &
Spa Resort

传承阿瓦丽高尔夫温泉度假村

Completion date: 2009
Location: Mauritius
Designer: Jean Marc Eynaud Architects
Photographer: Jean Marc Eynaud Architects
Area: 1,865 sqm

竣工时间：2009年
项目地点：毛里求斯
设计师：珍·马克·伊诺德建筑事务所
摄影师：珍·马克·伊诺德建筑事务所
项目面积：1,865平方米

Drawing on the fascinating cultural rainbow of Mauritius, the resort is located on the south coast of Mauritius on Le Domaine de Bel Ombre. Built on the "Domaine de Bel Ombre" set between crystalline lagoon and lush green hills, the "Heritage Awali" hotel is built in African tropical style with an ethnic style decoration. Extensive use of natural materials in the architecture (timber, stone, shingles, thatch and slates) together with extensive use of water bodies and lush landscaping helps create a very unique "Mauritian tropical ambiance".

The Main Building located on the left part of the site, is built around a huge swimming pool with elegant pyramid-

shaped roofs made of timber shingles. The water from the ponds and pool is ever present around the Restaurant, the Bar and the Lounges to give a strong tropical atmosphere to the place.

One hundred and fifty Deluxe Rooms, five Suites and a Presidential Villa complement the accommodation. The rooms are laid out in large courtyards all with sea views and wind protection. All rooms are decorated in a refined style inspired by chic ethnic themes which conjure up an atmosphere that is both tropical and African. Each of the rooms is spacious and comfortable and opens onto its own private and comfortable terrace, a stone's throw away from the two swimming pools or on the edge of the beach. The bathrooms reflect the latest in contemporary design, with their square basins and oval baths framed in black stone.

A Beach Restaurant and a swimming pool are located on the beachfront near the room clusters, with a relaxed atmosphere. Guests could enjoy toes in the sand, between the sea and the swimming pool.

A village-type tropical Spa is built at the rear of the bedrooms in the middle of a lush tropical garden. There are twenty treatment rooms, seven massage rooms of which three are double, two massage kiosks on the beach,

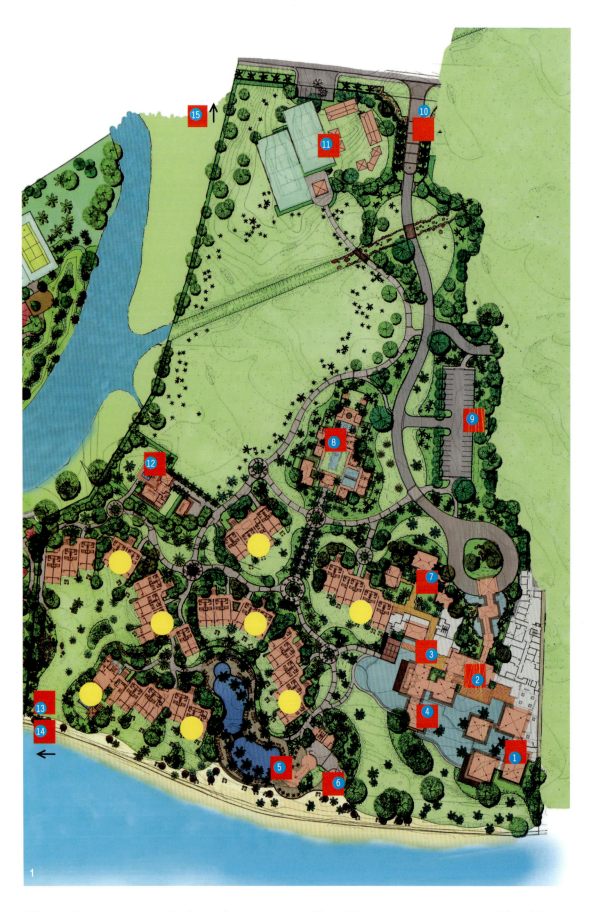

1. Balafon
2. Lobby, Reception, Guest Relation Desk, Library and Blue Earth Shop
3. Seven Colours Energy Cuisine
4. Zenzi Bar
5. Infinity Blue Restaurant and Blue Drift Bar
6. Diving centre and boathouse
7. Departure room/ Meeting room
8. Seven Colours Spa
9. Parking
10. Main Entrance
11. Umuzi Sports Centre/ Tennis, Beach Volley & Fitness Centre
12. Timomo & Friends Miniclub
13. Heritage Le Telfair Golf & Spa Resort
14. C Beach Club
15. Heritage Golf Club/ Le Chateau de Bel Ombre/ Frederica Nature Reserve/Heritage Villas

1. 木琴别墅
2. 大堂、前台、客户投诉台、图书室和蓝土商店
3. 七色能量美食餐厅
4. 禅思吧
5. 无尽之蓝餐厅和蓝色漂流酒吧
6. 潜水中心和船屋
7. 出发室/会议室
8. 七色水疗中心
9. 停车场
10. 主入口
11. 乌木兹运动中心/网球、沙滩排球和健身中心
12. 蒂默默和朋友迷你俱乐部
13. 高尔夫球场和水疗度假村
14. C海滩俱乐部
15. 传承高尔夫俱乐部/贝尔欧布莱城堡/弗雷德里卡自然保护区/传承别墅

Vichy showers, steam baths, a huge sauna with cold water pool, a pool with eddying currents, divans in kiosks especially provided for relaxation and one consulting room where tailor-made treatment programmes can be discussed.

1. Resort plan
2. Aerial view of the resort

1. 度假村平面图
2. 度假村鸟瞰图

度假村坐落在毛里求斯南海岸的贝尔欧布莱葡萄园，勾勒出毛里求斯那魅力十足的文化彩虹。度假村位于葡萄园清澈的湖泊和茂密的青山之间，具有非洲热带风情，充满了民族特色。建筑中大量自然材料（木材、石材、木瓦、茅草和石板）的运用与大量的水体和郁郁葱葱的景观相结合，共同营造出独特的"毛里求斯热带风情"。

坐落在场地左侧的主建筑环绕着一个巨大的游泳池而建，采用优雅的金字塔造型木瓦屋顶。池塘和水池的水环绕着餐厅、酒吧、休闲吧，营造出强烈的热带气氛。

度假村共有150套豪华客房、5套套房和一个总统别墅。客房分布在巨大的庭院之中，享有海景和防风保护。所有客房都以别致的民族主题进行了精致的装饰，形成了兼具热带和非洲风情的气氛。客房宽敞舒适，通往舒适的私人平台，距离游泳池和沙滩仅几步之遥。浴室反映了最新的现代设计，方形水盆和椭圆形浴缸采用了黑色岩石外框。

海滨餐厅和游泳池坐落在紧邻客房的沙滩上，散发出放松的氛围。宾客可以光着脚在大海和泳池之间的细沙上行走。

具有乡村风格的热带温泉水疗中心建在客房的后部，坐落在茂密的热带花园里。温泉水疗中心共有20间治疗室、7间按摩室（其中2间为双人按摩室）、2座海滩按摩亭、维希淋浴、蒸汽浴、带有冷水池的大型桑拿房、涡流水池、凉亭休息室和一间为客人量身定制治疗项目的咨询室。

1. Beach view
2. Infinity Blue Restaurant
3. Villa swimming pool
4. Restaurant exterior

1. 海滩
2. 无尽之蓝餐厅
3. 别墅游泳池
4. 餐厅外景

1. Seven Colours Spa
2. Seven Colours Spa
3. Pool view
4. Seven Colours Spa

1. 七色水疗中心
2. 七色水疗中心
3. 游泳池
4. 七色水疗中心

1

1. Zenzi Bar lounge
2. Seven Colours Energy Cuisine

1. 禅思吧
2. 七色能量美食餐厅

1. Villa guestroom
2. Deluxe room
3. Suite living room

1. 别墅客房
2. 豪华客房
3. 套房客厅

Banyan Tree Seychelles

塞舌尔悦榕庄度假村

Completion date: 2006
Location: Seychelles
Designer: Architrave Design and Planning Pte Ltd
Photographer: Banyan Tree
Area: 473,500 sqm

竣工时间：2006年
项目地点：塞舌尔
设计师：阿奇特拉夫设计规划公司
摄影师：悦榕庄集团
项目面积：473,500平方米

Nestled in Intendance Bay with spectacular views of the Indian Ocean, one of the world's most beautiful beaches, swaying palm trees and lush tropical forest, the Banyan Tree hotel and spa resort in Seychelles provides a rare taste of paradise. For the very best of Seychellois architecture, its 60 stunning pool villas combine contemporary, colonial and "plantation" décor; from high sloping ceilings, airy verandas and louvred doors to ethnic woven textiles. The eclectic villa interiors provide a true sense of place.

Hillside Pool Villa's inspired architecture and interior furnishings reflect the natural style of the Seychelles

through the use of indigenous materials. Around the panoramic environment, amidst the canopy of sheltering trees, privacy and tranquility are easily found.

An eclectic restaurant is with a homely, African ethnic feel to it. Unique Southeast Asian touches are incorporated into the service style and atmosphere of the restaurant, like the countries of Southeast Asia – multi-faceted, exotic and intriguing. The interior decorations have been selected piece by piece from all over Southeast Asia.

With blue skies and the stunning Indian Ocean at your doorstep, create your very own island fantasy in the Beachfront Spa Pool Villa. Local artist George Camille was commissioned to produce unique art pieces based on the legendary coco-de-mer, the world's largest nut, adding to the exotic splendour of these romantic villas. A delightful villa garden offers direct access to the silky soft beach.

Exuding a tropical old world charm, the Intendance Pool Villa is a classic example of traditional architecture. Each of the five villas boasts the finest in luxury with incomparable views of Intendance Bay and the Indian Ocean.

1. Exterior view of the resort
2. Presidential villa exterior view

1. 度假村外观
2. 总统套房外景

塞舌尔悦榕庄温泉度假村坐落在世界上最美的海滩之———英坦丹斯湾，俯瞰着印度洋的壮丽景色，周围环绕着高大的棕榈树和郁郁葱葱的热带森林，宛若天堂。度假村的60套泳池别墅完美地体现了塞舌尔的建筑风格，结合了现代、殖民地和种植园风格装饰：高耸的天花板、轻快的游廊、百叶窗门和民族风织物各具特色。折衷主义的别墅室内设计充满了真实感。

山坡泳池别墅富有灵感的建筑和室内装饰通过本地材料的运用反映了塞舌尔岛的自然风格。在高大的树木遮蔽下，人们能轻易地找到私密感和宁静感。

折衷主义风格餐厅拥有浓烈的家居感和非洲民族气息。独特的东南亚韵味被融入餐厅的服务风格和整体气氛之中，正如东南亚国家一样，多姿多彩而具有异国魅力。来自东南亚的室内装饰品经过了精挑细选。

蔚蓝的天空和迷人的印度洋近在咫尺，宾客可以在海滨温泉泳池别墅打造自己的梦幻之岛。当地艺术家乔治·卡米尔受邀以传奇的海椰子（世界上最大的坚果）来打造独特的艺术品，为这些浪漫的别墅增添了瑰丽的风情。宜人的别墅花园将直接通往丝般柔软的沙滩。

海湾泳池别墅流露出传统的热带风情，是传统建筑的典范。五座别墅以其无可匹敌的海湾海景而骄傲。

1. Au Jardin d'Epices	1. 失乐园
2. La Varangue	2. 瓦蓝格
3. Saffron Restaurant	3. 藏红花餐厅
4. Chez Lamar	4. 拉马尔之家
5. Beauty Salon	5. 美容沙龙
6. Board Room	6. 会议室
7. Gymnasium	7. 健身房
8. Helipad	8. 直升机停机坪
9. Library	9. 图书室
10. Lobby	10. 大堂
11. Spa Lobby	11. 水疗中心大堂
12. Tennis Court	12. 网球场
13. Banyan Tree Gallery	13. 悦榕庄画廊
14. Banyan Tree Spa Gallery	14. 悦榕庄水疗中心画廊
15. Presidential Villa	15. 总统别墅
16. Intendance Pool Villas	16. 管理泳池别墅
17. Beachfront Spa Pool Villas	17. 海滨水疗泳池别墅
18. Two-Bedroom Double Pool Villas	18. 两室双泳池别墅
19. Hillside Pool Villas	19. 山边泳池别墅
20. Banyan Tree Spa	20. 悦榕庄水疗中心
21. Pool Villas By-the-Rocks	21. 岩石边泳池别墅
22. Resort Pool	22. 度假村游泳池

1. Recreational facilities: the main pool
2. Resort plan
3. Intendance Pool Villa sun loungers
4. Intendance Pool Villa exterior view

1. 娱乐设施中的主游泳池
2. 度假村平面图
3. 高级泳池别墅日光浴床
4. 高级泳池别墅外部

1. Two-Bedroom Double Pool Villa exterior view
2. Sea and Stars Restaurant
3. Two-Bedroom Double Pool Villa outdoor lounge
4. Spa pavilion

1. 双卧双泳池别墅外观
2. 大海和星辰餐厅
3. 双卧双泳池别墅外部休息室
4. 水疗馆

1. Two-Bedroom Double Pool Villa guestroom
2. Two-Bedroom Double Pool Villa guestroom
3. Beachfront Spa Pool Villa bedroom

1. 双卧双泳池别墅客房
2. 双卧双泳池别墅客房
3. 海滨水疗泳池别墅卧室

Raffles Praslin Seychelles

塞舌尔普拉兰岛莱弗士度假村

Completion date: 2010
Location: Praslin, Seychelles
Designer: Wilson Associates Interior Design
Photographer: Raffles Hotels
Area: 1,335 sqm

竣工时间：2010年
项目地点：塞舌尔，普拉兰岛
设计师：威尔逊室内设计事务所
摄影师：莱弗士酒店集团
项目面积：1,335平方米

Located on the northeast tip of Praslin Island, the second largest island in the Seychelles, the Raffles Estates freestanding homes were built on pleasantly secluded. With 500 metres of direct beachfront land, spa and leisure facilities, restaurants, this resort will surely be a sought-after destination – private, relaxing, close to nature, and offering the heartfelt and gracious service for which Raffles is renowned. The sweet perfume of natural wood draped in elegant area rugs; bookshelves lined with knowledge and intrigue; beds dressed in sheer romance; a gourmet kitchen prepared to give discerning palettes a taste of something fresh and new; every room in your

Raffles home has mastered the delicate dances of style and taste.

All Raffles homes offer a private microcosm of paradise, marked by generous outdoor space and private pools, tended to by the discreet Raffles staff. Though Seychellois materials, art and influence can be found in every detail of your sanctuary, your senses will be inspired by discreet, yet state-of-the-art audio visual equipment.

As gentle as a Praslin breeze is the expert touch of your Raffles Amrita Spa. Here, treatments that both relax and acclimatise the body to the power and rhythm of a tropical world are necessary rituals of daily life. From ancient ceremonies involving healing plants, to secluded areas overlooking the ocean, to expansive changing rooms, the resort offers scented rain showers and soothing steam. Your Raffles Amrita Spa shares the secrets to living well, lavishing you in layer upon layer of earthly pleasures, until your spirit feels as lustrous and fine as a South Sea pearl.

莱弗士度假村坐落在塞舌尔第二大岛——普拉兰岛的东北角。度假村拥有500米长的海岸线，设有温泉休闲设施和餐厅，是梦想中的度假胜地——私密、放松、亲近自然，为宾客提供莱弗士著名的贴心而精致的服务。散发着香气的天然木地板上面覆盖着优雅的装饰地毯；书架里满是知识和复杂的情节；大床浪漫迷人；美食厨房提供新鲜的精致美食；莱弗士度假村的每个房间都独具品味。

每间莱弗士别墅都是一个微缩的私人天堂，以宽敞的露天空间和私人泳池为特色，贴心的莱弗士员工将提供完美的服务。塞舌尔特有的材料、艺术和风情将体现在别墅的方方面面，最新的影音设备将为宾客带来非凡的感官体验。

莱弗士仙露温泉水疗中心将提供如普拉兰和风般柔和的专业服务。在这里，放松并让身心适应热带岛屿气候的温泉理疗是每日必不可少的。在采用治愈植物的古老仪式、远眺海洋的隐蔽区域乃至宽敞的更衣室，度假村都提供淋浴和蒸汽浴。仙露温泉水疗中心运用健康的秘诀让人感受尘世的愉悦，直到灵魂得到净化，如东海明珠般闪亮、精致。

1. Resort plan
2. Exterior view of the villa
3. Arrival hall

1. 度假村平面图
2. 别墅外观
3. 到达大厅

1. Libreri	1. 李布莱里亭
2. Arrival pavilion	2. 到达大厅
3. Buggy pick-up & drop off	3. 巴奇上客/下客区
4. Boutique	4. 精品酒店
5. Main road	5. 主路
6. Lobby lounge	6. 大堂休息厅
Reception & concierge	前台和门房
7. Losean restaurant & Deli	7. 罗西恩餐厅和食品店
8. Takamaka terrace lounge	8. 塔卡玛卡休息露台
9. Danzil bar & lounge	9. 丹泽尔酒吧
10. Anse Boudin river	10. 安斯·布顿河
11. Raffles spa	11. 莱弗士水疗中心
Fitness centre	健身中心
Great room	公共休息室
Movement pavilion	运动亭
Beauty salon	美容沙龙
12. Anse Takamaka beach	12. 安斯·布顿海滩
13. Boat house	13. 船屋
14. Main pool	14. 主游泳池
15. Walkway	15. 走道
16. The sugar palm club	16. 糖椰子树俱乐部
17. Curieuse restaurant & pool bar	17. 好奇餐厅和泳池吧

1. Ocean view villa plunge pool
2. Spa-outdoor pavilion
3. Spa couple suite

1. 海景别墅泳池
2. 户外水疗亭
3. 水疗情侣套房

1. Curieuse Restaurant
2. Losean Restaurant – dining
3. Losean Restaurant – breakfast

1. 好奇餐厅
2. 罗西安餐厅——就餐区
3. 罗西安餐厅——早餐区

1. The Sugar Palm Club
2. Raffles Suite Villa – master bedroom
3. Hillside View Pool Villa – twin room

1. 糖心椰子俱乐部
2. 莱弗士套房别墅——主卧
3. 山景泳池别墅——双人房

1. Ocean view pool villa - bedroom
2. Bathroom
3. Royal Suite Villa living room
4. Bathroom

1. 海景泳池别墅客房
2. 浴室
3. 皇家套房别墅客厅
4. 浴室

Maia Luxury Resort & Spa

美雅奢华水疗度假村

Completion date: 2006
Location: Mahé, Seychelles
Designer: Bill Bensley
Photographer: Maia Luxury Resort & Spa
Area: 48, 562 sqm

竣工时间：2006年
项目地点：塞舌尔，马埃岛
设计师：比尔·本斯利
摄影师：美雅奢华水疗度假村
项目面积：48, 562平方米

When it is time to pause your frenetic pace and rediscover senses dulled by the incessant demands of life, there is a place called Maia Luxury Resort & Spa – offering something more than just breathtaking views, unsurpassed service and astonishing accommodation.

Discover peace and privacy on the private peninsula on the island of Mahé in the Seychelles where the sands of the secluded Anse Louis beach are washed by the timeless tides of the warm Indian Ocean. It is a palm-filled private paradise of lush tropical gardens with a policy of absolute discretion that ensures complete freedom.

There are 30 spacious one-bedroom, air-conditioned, top hat-thatched villas with uninterrupted ocean views. Each villa is approximately 250 square metres in size and has a spacious gazebo area, adjacent bar, an oversized day-bed, an infinity pool and outdoor and sunken double bath tubs. There are three Ocean View Villas, eight Ocean Front Villas, eleven Ocean Panoramic Villas and eight Maia Signature Villas. All the villas are identical, all with the same interior design; they vary only in their location.

The Maia Spa by La Prairie offers world-class therapy and beauty treatments using La Prairie products and the highly-trained staff can also offer accompanied yoga, self-Shiatsu and Qi Gong sessions.

1. Bar
2. Dining
3. Sun deck
4. Bedroom
5. Pool
6. Bath
7. Bathroom

1. 酒吧
2. 餐厅
3. 日光浴平台
4. 卧室
5. 游泳池
6. 浴缸
7. 浴室

当你想停下疯狂的脚步，重新找回被生活变得迟钝的感官，美雅奢华水疗度假村正是你的天堂。除了惊艳的美景、非凡的服务和优异的住宿之外，美雅奢华水疗度假村将为你带来更多的精神享受。

在这座位于塞舌尔马埃岛上的私人半岛上，隐蔽的路易海湾沙滩上的细沙被温暖的印度洋不断冲洗。宾客将体会到前所未有的平和感和私密感。在这个郁郁葱葱的热带私人天堂，宾客享有绝对酌情决定权，拥有完全的自由。

30套宽敞的单间空调茅草顶别墅享有一览无余的海洋美景。每座别墅面积约250平方米，拥有宽敞的露台、小吧台、超大的日光浴床、无边界泳池和露天下沉式双人浴缸。度假村共有3座海景别墅、8座朝海别墅、11座海洋全景别墅和8座玛雅特色别墅。所有别墅都拥有完全相同的室内设计，仅仅在地理位置上略有不同。

大草原美雅水疗中心采用大草原产品提供世界级疗法和美容理疗服务，训练有素的员工还将提供相应的瑜伽、日式指压和气功教程。

1. Ocean Front (Beach) Villa roof plan
2. Ocean Front (Beach) Villa ground floor plan
3. Overall view of the resort
4. View of main swimming pool

1. 海滨别墅屋顶平面图
2. 海滨别墅一层平面图
3. 度假村全景图
4. 主游泳池

1. View from the villa
2. View from the villa
3. Spa pavilion
4. Bath pavilion

1. 别墅
2. 别墅
3. 水疗馆
4. 浴缸亭

1. Dining pavilion
2. Dining pavilion
3. View from the villa

1. 就餐亭
2. 就餐亭
3. 别墅

3

1. Villa bedroom
2. Villa bedroom
3. Villa bedroom

1. 别墅卧室
2. 别墅卧室
3. 别墅卧室

Elounda Beach Hotel

伊罗达海滨酒店

Completion date: 2008 (renovation)
Location: Elounda, Crete, Greece
Designer: Davide Macullo Architects (Yachting Villas)
Photographer: Elounda Beach Hotel
Area: 161,874 sqm

竣工时间：2008年（翻新）
项目地点：伊罗达，希腊，克里特岛
设计师：大卫·马库洛建筑事务所（游艇别墅区）
摄影师：伊罗达海滨酒店
项目面积：161,874平方米

Elounda Beach Hotel is a five-star luxurious resort. Featuring 249 guestrooms customers will have the chance to opt for their favourite room. The design and decoration of all rooms stem from the traditional Greek Island architecture combined with a discreet touch of luxury. The interior unveils thoughtful details in every room, fitted with handmade furniture and luxurious marble bathrooms, private balcony, terrace or garden, which is tastefully laid out providing a natural kaleidoscope of colours.

Common to all rooms is the range of high-tech gadgetry (indeed, even finding the desired light switch can be

arduous) as well as the prevailing maritime design theme, which highlights the hotel's Greek island location. White-washed exteriors with blue-painted doors contain spacious interiors featuring rich polished wood floors bathed in deep green or blue marble. Rooms are all quite clean and well-maintained.

Although it is not the biggest category of room size wise, the unique seaside bungalow – constructed at a time when Greece still allowed rooms to be built close to the water – is a favourite with many. These rooms follow a simple straight-on entry design, with bathroom (containing sunken Jacuzzi and hamam/steam shower as well as fitness equipment and internal speakers) located behind a large, high-ceilinged bedroom with views of the sea through enormous floor-to-ceiling windows. Open the door and it's only a ten-second walk from one's private enclosed veranda down to the bay and small ladder leading into it. The glamorous Elounda Beach Resort, positioned on the tranquil northeastern coast of Crete, offers the most discerning guests the ultimate in comfort and privacy with the new ultra luxurious Yachting Club Villas. The cutting-edge design and high-tech facilities provide the ultimate waterfront paradise to unwind and relax in style. The spa, characterised by the use of "light" materials such

1. Main pool surrounding area on the east side of the resort
2. Yachting club
3. Blue Lagoon Polynesian & Sushi Bar

1. 度假村东部主泳池周边地区
2. 游艇俱乐部
3. 蓝色潟湖玻璃尼西亚与寿司酒吧

as glass and translucent panels for walls and ceilings, creates a calming visual neutrality and is gently animated by the effects of reflection, while the surrounding landscape dominates the space and sheds its colours.

The essentiality of the materials and the graphic sign of the "leaves" pattern, contribute to stretch the perspective throughout an atmosphere of pastel colours and lightness, emphasised by the use of indirect artificial lighting at night.

1. Yachting suites
2. Plan of the conference rooms
3. Spa centre
4. Indoor pool of spa centre

1. 游艇套房
2. 会议区平面图
3. 水疗中心
4. 水疗中心的室内游泳池

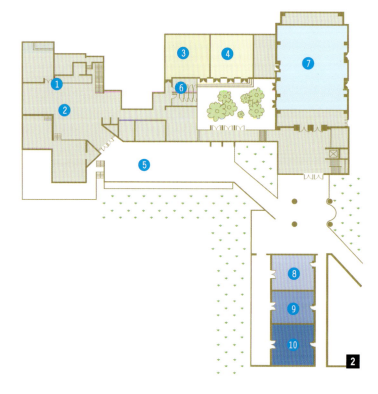

1. Reception
2. Lobby
3. Cronos A
4. Cronos B
5. Courtyard
6. WC
7. Zeus
8. Mars
9. Venus
10. Itios

1. 前台
2. 大堂
3. 克洛诺斯A房
4. 克洛诺斯B房
5. 庭院
6. 洗手间
7. 宙斯房
8. 战神房
9. 维纳斯房
10. 伊蒂欧斯房

伊罗达海滨酒店是一个奢华的五星级度假酒店。酒店拥有249套客房，可供宾客任意挑选。客房的设计和装饰灵感都来源于希腊岛传统建筑，并谨慎地添加了奢华感。每套客房都展现了精致的细节，配以手工打造的家具和奢华的大理石浴室、私人阳台、露台或花园（其巧妙的布局形成了自然色彩的万花筒）。

每间客房内装饰着高科技小电器（连开关都经过精挑细选），辅以流行的海洋设计主题，凸显了酒店的希腊风情。白色的建筑外观和蓝色的大门内部是宽敞的室内，以抛光木地板和深绿、蓝色大理石为特色。所有房间都洁净而舒适。

尽管并不是面积最大的客房，独具特色的海滨小屋还是深受宾客欢迎。这些小屋建于希腊还允许建造水边建筑的时期，设计十分简洁。巨大的海景卧室配有落地窗，后面是配有下沉式极可意按摩浴缸和蒸汽浴室、健身设备和内部对讲机的浴室。打开大门，仅需十秒钟便从私人封闭游廊通过小梯子走到海湾。

迷人的伊罗达海滨度假村位于克里特岛宁静的东北海岸，其奢华的游艇俱乐部别墅将为眼光独到的宾客提供终极的舒适和隐秘度假体验。前沿的设计和高科技设施共同打造了轻松愉快的滨水天堂。

度假村的温泉水疗中心以其独特的"轻质材料"而闻名，例如玻璃、半透明墙壁板和天花板营造出迷人的视觉中立感。倒映的效果让整个氛围活跃起来，而周边的风景则为空间增添了色彩。

这些材料的本质和树叶图案让人们的视野穿越浅淡而轻盈的氛围，而夜晚的间接照明则凸显了这一效果。

1. Sea majesty of Yachting Villa
2. Spa relaxation area
3. Spa relaxation area
4. Sea Pearl Yachting Villa
5. Island Suite bedroom

1. 游艇别墅纵览海景
2. 水疗休息区
3. 水疗休息区
4. 海洋珍珠游艇别墅
5. 小岛套房卧室

Cap Juluca
卡普朱卢卡度假村

Completion date: 2008
Location: Anguilla, British West Indies
Designer: Paul Duesing
Photographer: Cap Juluca
Area: 724,387 sqm

竣工时间：2008年
项目地点：英属西印度群岛，安圭拉岛
设计师：保罗·德赛因
摄影师：卡普朱卢卡度假村
项目面积：724,387平方米

Playing off the cultural magnificence of the Caribbean and the subtleties of the natural beauty of Anguilla, the designer's goal for the décor of Cap Juluca is to shine in the day with colourful backsplashes amid airy, illuminated compositions and at night with dramatic lighting and shadowing effects.

Through commissioned artistry from Moroccan craftsmen, Duesing was able to utilise custom aesthetic pieces to create a parallel between the interior framework and the alluring Moorish architecture. The Main House has been transformed into a surreal "gathering place" with limestone flooring, distinctive handmade rugs, centuries-

old artifacts, antique wooden game tables and intricate serving trolleys.

Under the dome of the expanded open-air lobby hangs a showpiece 900-pound chandelier constructed of punched tin dipped in brass – intentionally not varnished to create a "tarnished" aquamarine patina effect over time. The chandelier consists of 94 decorative metal balls strategically placed to refract light and simulate the twinkling of an Anguillian starry night.

While dining in one of Cap Juluca's three restaurants, guests are dazzled by Duesing's handiwork, as new designs complement the revived dining concepts. Serving upscale "Eurobbean" fare, Pimms is set on a coral outcrop where spray from the surf splashing against the rocks creates an aerial ballet for amorous diners. With a slight nod to British tradition, the design is "understated sophistication" with light Pistachio hues and crisp whites accented only with nightscape views of the twinkling lights of St. Maarten. The Pan-Asian concept restaurant, Spice, overlooks the lighted Moorish domes, turrets and parapets of the resort. Its provocative atmosphere comes alive at night, with antiqued reds and oranges accented by impressive hand-blown glass from Mexico, Buddha statues and hundreds of candles reflecting off the silver-

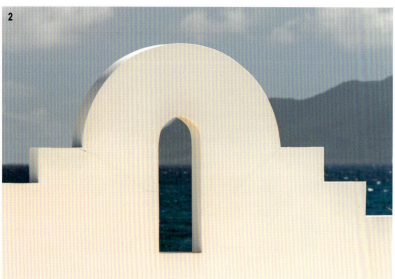

leafed ceiling varnished with a fiery hued glaze.

The open-air beachfront restaurant, Blue, is designed to feel casually elegant with a Caribbean twist, featuring teak wood flooring painted white and pale azure interiors, accented with punchy raspberry, lime and orange chairs. Blue delivers an array of international cuisine and panoramic views of the turquoise waters of the Caribbean Sea lapping sweeping white sands.

Guest quarters are elegantly appointed, incorporating vibrant hues and decidedly modern yet understated amenities. From the five-room villas to the junior suites, Duesing complements the intriguing architecture with his use of authentic Moroccon rugs, detailed stone patio tables, aesthetically intriguing mirrors and enormous candle lanterns. Unparallelled ocean views serve as the breathtaking artwork.

In Duesing's new room, the bed is the centrepiece, veiled in a sheer, elegant canopy from South Africa and covered with white Frette linens and white hand-sewn bedspreads. Behind the bed, an inlaid arch featuring an intricate mosaic design provides an element of drama and romance to the focal point. As Duesing believes a bathroom should be a lady's "liquid temple", the washrooms become just that with travertine marble, lavish light fixtures, substantial soaking tubs, separate enhanced showers, double sinks and large dressing tables with cushioned banquettes.

卡普朱卢卡度假村结合了加勒比的文化和安圭拉岛的自然美景，设计师的目标是让度假村在白天显得轻快而多姿多彩，在夜晚展示戏剧化的光影效果。

设计师通过摩洛哥工匠精湛的工艺，使用定制艺术品在室内设计和摩尔式建筑之间实现了平衡。主屋被改造成为一个超现实"集会空间"，装饰着石灰岩地板、独特的手工地毯、古老的艺术品、古董木制游戏桌和精致的服务推车。

在露天大堂巨大的穹顶下，一盏900磅的锡铜吊灯特意未经刷漆，营造出古旧的铜锈效果。吊灯由94个装饰性金属球巧妙布置而成，折射着光线并为安圭拉岛的繁星之夜增添了点点星光。

卡普朱卢卡度假村共有三家餐厅，餐厅里的手工艺品令宾客眼花缭乱，新设计完善了鲜活的餐饮理念。皮姆斯酒店坐落在珊瑚岩上，提供高档"欧式加勒比"美食，海浪的泡沫飞溅到岩石上，为热情的就餐者呈现了空中芭蕾。设计中添加了少许英伦传统，淡草绿色和纯粹的白色所呈现的低调的奢华在星光闪闪的夜景中格外突出。泛太平洋概念餐厅——香料餐厅俯瞰着度假村光亮的摩尔穹顶、角楼和矮墙。餐厅挑逗的氛围在夜间达到了极致，古旧的红色和橙色在墨西哥人工吹制的玻璃、佛像和成百上千的烛火中尤为突出，银箔天花板涂有火红的釉面。

露天海滩餐厅——蓝色餐厅的设计休闲优雅，充满了加勒比风情，以白色的柚木地板和浅蓝色的室内设计为特色，点缀着浓烈的紫红色、黄绿色和橙色的座椅。蓝色餐厅提供各种国际美食，宾客们在就餐时可以享受加勒比海碧蓝的海水静静拍打白沙的美景。

客房区设计优雅，充满活力的色调和直接、现代而又低调的设施让人愉悦。从五居室别墅到小型套房，设计师在迷人的建筑中融入了摩洛哥地毯、精致的石台餐桌、美观

的镜子和巨大的提灯。无与伦比的海景则是最美的艺术品。

在德赛因设计的新卧室内，床是中心。来自南非的轻薄高雅的华盖、白色亚麻织物和白色的手工缝制床单装饰着大床。除了床铺以外，镶嵌拱门上复杂的马赛克设计为室内提供了浓烈的元素和浪漫感。因为德赛因认为，浴室是女士的"水神殿"，洗手间内以石灰华大理石、丰富的灯饰、坚固的浴缸、独立淋浴、双式水槽和巨大的梳妆台为特色。

1. Sand sculpture	1. 沙雕
2. Architecture Arch ocean view	2. 建筑拱门海景
3. Architecture Arch St Martin	3. 建筑拱门和圣马丁岛
4. Architecture Dome	4. 建筑穹顶
5. Architecture Maundays Bay	5. 建筑和芒茨斯海湾
6. Architecture Oscar Farmer Lines	6. 建筑和奥斯卡农夫航线
7. Pool Villa columns	7. 泳池别墅廊柱
8. Architecture top lines	8. 建筑顶部

1. The wedding lawn

2. Tennis court #2

   Tennis court #3

   Pro shop

   Pavilion

3. Herb garden waling trail

4. Croquet lawn

5. Blue

   Bule patio bar

Fresh water pool

Beach pavilion

6. Green house

7. Tennis court#1

8. The main house

   Guest reception

   Concierge

   The library (east & west)

   Zemi

Cap j boutique

Concierge

Manundays club

Under the dome

The private screening room

The wellness centre

9. Private Pool Villas

1. 婚礼草坪

2. 2号网球场

   3号网球场

   专卖店

   凉亭

3. 植物园横档小径

4. 门球场

5. 蓝色

   蓝色中庭吧

淡水泳池

海滨亭

6. 绿色房

7. 1号网球场

8. 主屋

   客房接待处

   门房

   图书室（东&西）

   泽米餐厅

J精品店

门房

玛农戴斯俱乐部

穹顶下方

私人放映厅

健身中心

9. 私人泳池别墅

1. Temenos golf course
2. Blue Café
3. Resort plan

1. 坦密诺斯高尔夫球场
2. 蓝色咖啡厅
3. 度假村平面图

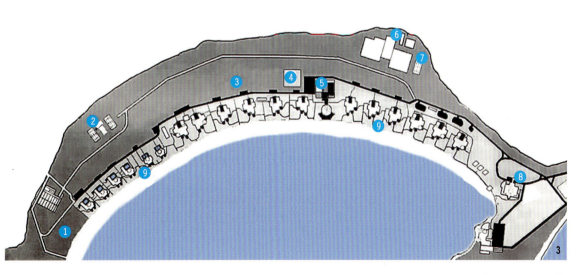

1. Beach Maundays Club
2. Dining Spice Restaurant
3. Under Dome dining
4. Under Dome dining

1. 海滨芒戴斯俱乐部
2. 风味餐厅
3. 穹顶餐厅
4. 穹顶餐厅

1. Maundays Bar
2. Pimms Wine Room
3. Flights Spice Restaurant

1. 芒戴斯酒吧
2. 皮姆斯品酒室
3. 飞香餐厅

1. Pool Villa exterior
2. Bathroom
3. Pool Villa living room
4. Pool Villa bathroom

1. 泳池别墅外观
2. 浴室
3. 泳池别墅客厅
4. 泳池别墅浴室

1. Library
2. Poll Villa living room
3. Suite bedroom
4. Pool Suite bathroom

1. 图书室
2. 泳池别墅客厅
3. 套房卧室
4. 泳池套房的浴室

Lindian Village Hotel

林甸乡村酒店

Completion date: 2009
Location: Island of Rhodes, Greece
Designer: MKV Design
Photographer: LHW
Area: 8,000 sqm

竣工时间：2009年
项目地点：希腊，罗德岛
设计师：MKV设计
摄影师：LHW
项目面积：8,000平方米

Discover heaven on earth in this jewel amongst luxury resorts in Rhodes Greece. Unique amongst luxury resorts, the picturesque Lindian Village in located on the breathtaking, turquoise Lardos Beach. Lindian Village is a deluxe resort, designed to resemble a Greek village, with a natural landscape and contemporary Greek character. The resort is in harmony with the Rhodian multi-cultural character where all different influences blend together in a perfect mix.

The River Passage luxury suites in Rhodes create a magical mix of deep blues and enchanting whites, while the Ottoman Gardens luxury suites in Rhodes are truly seductive with their stylish oriental hints and colours.

All suites offer the luxury of private pools and Jacuzzi. The opulent Mediterraneo Classic rooms are surrounded by heavenly scented bougainvilleas and a cornucopia of flowers.

Dressed in deep blues and stunning whites, The Suites are especially designed for those who seek complete privacy in absolute style. Set up amidst their own lush garden with private swimming pool and lovely sun deck, they guarantee seclusion wrapped in total luxury. Their characteristic feature is a four-poster bed, while their bathrooms comprise of his and her washing areas. All Suites include a lavish living room area and can accommodate up to three people.

Stylish Aegean-style rooms offer superb terraces nestled amidst lush greenery, luxurious Junior Suites boast living rooms, private terraces and an open-air Jacuzzi, whilst the Suites offer complete privacy with lavish living rooms, opulent four-poster beds and a private garden with sun deck and swimming pool.

Choose from a selection of fabulous gourmet restaurants, the cosmopolitan International Restaurant features simple yet refined cuisine, superbly prepared and served with flair and the Greek Restaurant offers dishes from mainland Greece and the Aegean Islands in a cool contemporary atmosphere.

1. Ottoman gardens
2. Pool
3. Resort plan

1. 土耳其花园
2. 泳池
3. 度假村平面图

1. Reception lobby

2. Almantes Main Restaurant

3. Mr. Danton Thai Restaurant

4. Bassil Greek Restaurant

5. La piazza coffee bar

6. Mini market

7. Main square

8. Tennis court

9. Chapel

10. LV Spa

11. Open-air gym

12. Swimming pool complex

13. Children swimming pool

14. Kohilo all-day pool restaurant

15. Pebble pool bar

16. Beach

17. River Passage Suites & Junior Suites

18. Ottoman gardens

19. Mediterraneo classic rooms

20. Astroscopus fish & oyster restaurant

1. 前台大堂

2. 阿尔曼特斯主餐厅

3. 丹顿先生泰式餐厅

4. 巴希尔希腊餐厅

5. 比萨咖啡吧

6. 迷你市场

7. 主广场

8. 网球场

9. 礼拜堂

10. LV水疗中心

11. 露天健身中心

12. 游泳池

13. 儿童游泳池

14. 谷希罗全日泳池餐厅

15. 卵石泳池吧

16. 海滩

17. 河畔套房和普通套房

18. 土耳其花园

19. 地中海经典客房

20. 阿斯特罗斯库珀斯海鲜餐厅

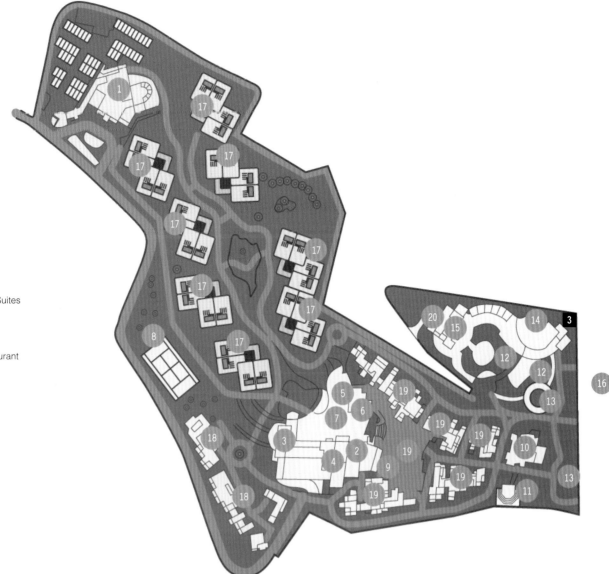

这座位于希腊罗德岛的奢华度假村就是人间天堂。风景如画的林甸度假村在众多奢华度假村中脱颖而出，坐落在碧蓝色的劳尔多斯海滩。林甸是一个豪华度假村，其设计与希腊村庄类似，享有自然景观和现代希腊特色。度假村与罗德岛的多重文化氛围相互交融，各种文化完美地结合在一起。

河道奢华套房神奇地结合了深蓝色和迷人的白色，而土耳其花园套房则以其独特的东方韵味和色彩吸引着人们。所有套房都配有奢华的私人泳池和极可意按摩浴缸。宽敞的地中海经典客房四周环绕着芳香的九重葛和巨大的花坛。

套房以深邃的蓝色和耀眼的白色为主调，专为追求时尚和隐私的人们设计。郁郁葱葱的私人花园里设有私人游泳池和可人的日光浴平台，以极致的奢华包围着桃源胜地。套房以四柱床为特色，其浴室分设男女洗浴区。所有套房都设有一个宽敞的起居区，最多能容纳三个人。

时尚的爱琴海风格客房提供了绿树环绕的平台；奢华的小型套房以起居室、私人平台和露天极可意按摩浴缸为特色；套房则提供了绝对的隐私，以其宽敞的起居室、巨大的四柱床和带有日光浴平台和泳池的私人花园而著称。

度假村的国际化餐厅提供简单而精致的美食，而希腊餐厅则提供希腊大陆和爱琴群岛的美食，拥有清爽的现代风格。

1. Basil reception
2. Beach
3. Mediterraneo classic room outside
4. Kohilo all-day pool restaurant
5. Basil Greek Restaurant outside
6. Basil Greek Restaurant inside

1. 巴兹尔招待会
2. 海滩
3. 地中海经典客房露天区
4. 谷希罗全日泳池餐厅
5. 巴兹尔希腊餐厅露天区
6. 巴兹尔希腊餐厅室内

1. Lobby
2. Almantes Restaurant
3. Heleni room

1. 大堂
2. 阿尔曼特斯餐厅
3. 海勒尼房

1. River Passage Suite outside
2. Sea spirit disco bar
3. Spa indoor pool
4. River Passage Pool Suite outside

1. 河畔走廊套房露天区
2. 海洋精神迪吧
3. 水疗中心的室内游泳池
4. 河畔走廊泳池套房露天区

1. Mediterraneo classic honeymoon room
2. Mediterraneo classic room
3. Mediterraneo classic room
4. Suite bedroom
5. Ottoman Pool Suite bedroom

1. 地中海经典蜜月房
2. 地中海经典房
3. 地中海经典房
4. 套房卧室
5. 土耳其泳池套房卧室

1. Pool Suite living room
2. River Passage Pool Suite bedroom
3. Ottoman Pool Suite bedroom
4. Ottoman Pool Suite bathroom

1. 泳池套房客厅
2. 河畔走廊泳池套房卧室
3. 土耳其泳池套房卧室
4. 土耳其泳池套房浴室

Turnberry, A Luxury Collection Resort, Scotland

坦伯利度假村

Completion date: 2009 (renovation)
Location: Ayrshire, Scotland
Designer: James Miller, Mary Fox Linton
Photographer: Turnberry Resort
Area: 3,237,485 sqm

竣工时间：2009年（翻新）
项目地点：苏格兰，埃尔郡
设计师：詹姆斯·米勒、玛丽·福克斯·林顿
摄影师：坦伯利度假村
项目面积：3,237,485平方米

Fabled white walls rise up to a thousand red tiles. Turnberry, A Luxury Collection Resort, Scotland, one of the most enchanting hotels in Scotland, stands sentinel above historic Ayrshire lands. Turnberry Resort offers spectacular scenery, two championship golf courses, the Ailsa and the Kintyre, the nine-hole Arran beginners course and the acclaimed Colin Montgomerie Links Golf Academy.

Turnberry has six restaurants, bars and lounges – the signature restaurant 1906, the relaxed Duel in the Sun Bar, The Ailsa Bar & Lounge, The James Miller Room and The Grand Tea Lounge, inspired by the hotel's

original tea lounge in the Edwardian era. In the heart of the Turnberry kitchens is the ultimate dining experience, The Turnberry Chef's Table. For golfers, Turnberry also boasts the Tappie Toorie restaurant, which is situated in the Clubhouse.

Featuring an inviting 20-metre indoor pool, eleven individual treatment rooms, fitness studio, heat experiences, pool-side Jacuzzi, hair salon and children's pool and offering a range of treatments by ESPA including reflexology and hot stone massage, the Spa at Turnberry is the perfect place to relax and reinvigorate. There are a total of 149 rooms, including four Specialty Suites and four eight-bedroom lodges. Eight two bedroom Lands of Turnberry luxury self-catering apartments are also located in the heart of the resort. The beautiful ocean-view rooms capture the true spirit of Turnberry. Soft fabrics, subtle colours and honeyed wooden flooring give each room a natural purity, at one with its setting on the Scottish coast. Generous picture windows flood the room with natural light, offering evocative views over slate seas, mystical islands and green links. The Deluxe Rooms offer a perfect blend of Edwardian tradition and contemporary design. There are views over the rolling Turnberry estate or

1. Main hotel
2. The Spa
3. Clubhouse
4. The Colin Montgomerie
   Links Golf Academy
5. Trout fishing
6. Lands of Turnberry outdoor pursuits
7. Lodges

1. 酒店主体区
2. 水疗区
3. 俱乐部
4. 科林·蒙哥马利高尔夫学会
5. 垂钓区
6. 坦伯利度假村户外活动区
7. 客房区

1. Resort plan
2. Local area: Culzean castle
3. Local area: Brig O Doon
4. Local area: Burns Statue

1. 度假村平面图
2. 当地的卡尔津城堡
3. 当地的杜恩河
4. 当地的伯恩斯雕像

the new arrivals in the courtyard below. The Classic Ocean-View Rooms, located above The Spa at Turnberry, offer breathtaking views over the links golf courses, the Irish Sea, the Ailsa Craig and Isle of Arran.

1. The lodges at Turnberry
2. Local area: Dunure castle
3. Local Alloway Burns Monument

1. 坦伯利民居
2. 当地的杜奴力城堡
3. 当地的阿洛韦·伯恩斯纪念塔

白墙红砖交互掩映。作为苏格兰最迷人大酒店之一，坦伯利度假村高高耸立在古老的埃尔郡。坦伯利度假村拥有壮丽的景色、两座锦标赛高尔夫球场、艾尔莎和琴泰半岛、九洞阿蓝岛新手球场和著名的克林·蒙哥马利海湾高尔夫学校。

坦伯利度假村拥有六家餐厅、酒吧和休闲吧——特色餐厅1906、太阳吧内的休闲决斗餐厅、艾尔莎酒吧和休闲吧、詹姆斯·米勒餐厅和宏大茶室（其设计灵感源自爱德华时代酒店的茶室）。坦伯利厨房的核心是终极餐饮体验——坦伯利大厨餐桌。坦伯利还在俱乐部内专门为高尔夫球手提供了他比·突利餐厅。

坦伯利度假村的温泉水疗中心以20米长的室内泳池、11间独立治疗室、健身工作室、高温浴室、池边极可意按摩浴缸、美发沙龙和儿童泳池为特色，为宾客提供全套的温泉治疗（包括反射疗法和热石按摩），是获得充分放松、重获新生的理想之地。

度假村共有149套客房，其中包括4套特别套房和4套八卧室的别墅。同时坦伯利度假村中心还有8座自备餐饮的双人间奢华公寓。美丽的海景房深得坦伯利的精髓。柔软的织物、精巧的色彩和蜜色木地板为每间房间都带来了自然纯净感，与苏格兰海岸景色相得益彰。大片的落地窗让室内洒满阳光，提供蔚蓝色的浩瀚海洋、神秘的岛屿和绿色植被的美景。豪华套房完美地结合了爱德华时期的传统风格和现代设计，俯瞰着坦伯利起伏的景色或是庭院里新到的宾客。经典海景客房位于坦伯利温泉水疗中心上方，远眺着高尔夫球场、爱尔兰海、艾尔莎克雷格岛和阿蓝岛的美景。

1. Hotel exterior
2. Burns Alloway cottage
3. The Turnberry Clubhouse

1. 酒店外观
2. 阿洛韦·伯恩斯故居
3. 坦伯利俱乐部

1. The Gallery room
2. Brucc's Room – meeting room
3. 1906 Restaurant
4. The Caledonia Suite

1. 宴会厅
2. 布鲁斯房——会议室
3. 1906餐厅
4. 喀里多尼亚套房

1. Brown's Room – private dining
2. The Centenary Room – private function space
3. Grand tea lounge
4. The Tappie Toorie
5. Deluxe ocean-view room
6. Suite detail

1. 布朗房——私人用餐雅间
2. 圣塔利房——私人功能用餐间
3. 大茶室
4. 塔皮杜里餐厅
5. 豪华海景房
6. 套房细部

1. Watson Suite
2. Classic ocean-view room
3. Norman Suite
4. Classic bedroom

1. 沃森套房
2. 经典海景房
3. 诺曼套房
4. 古典卧室

1. Deluxe resort view room
2. Deluxe ocean-view room
3. Family room
4. Deluxe bathroom
5. Deluxe bathroom

1. 豪华度假房
2. 豪华海景房
3. 家庭套房卧室
4. 豪华浴室
5. 豪华浴室

Fairmont Southampton

南安普顿费尔蒙特度假村

Completion date: 2006
Location: Southampton, UK
Designer: Frank Nicholson
Photographer: Fairmont Hotels & Resorts
Area: 400,000 sqm

竣工时间：2006年
项目地点：英国，南安普顿
设计师：弗兰克·尼克尔森
摄影师：费尔蒙特酒店度假村集团
项目面积：400,000平方米

The Fairmont Southampton is decorated in a tastefully conservative style with rare but usually welcome touches of glitziness and razzmatazz, always with well-upholstered furnishings and a sense of airy spaciousness. Baronial staircases connect the public rooms, situated on three floors. The plush guestrooms are arranged in wings that radiate, more or less, symmetrically from a central core. This design gives each luxurious room a private veranda with a sweeping view of the water. Rooms are spacious, each with a larger-than-expected balcony, and either one king or two double beds. For those who can afford it, the choicest accommodations are on the

Fairmont Gold Floor, a hotel-within-a-hotel, pampering its guests in ultimate luxury on the top floor, offering an array of services from shoeshines to private check-ins. The Gold Floor offers complimentary continental breakfast, newspapers, and the use of a fax and VCR.

One- and Two- Bedroom Bermuda Hotel Suites offer a separate living room area and are equipped with a second TV and mini bar. Luxuriously furnished with overstuffed sofa, chairs and original art work inspired by Bermuda's soft pastels draped in a flowing sea of turquoise. These Bermuda Suites feature two private balconies with two fully-appointed marble bathrooms, walk-in closets, a desk and a second telephone line for High Speed Internet access and for convenience, a cordless telephone. One- and Two-Bedroom Bermuda Suites are available on all floors. For a romantic dinner, reserve a table at the elegant Newport Room designed to evoke a luxury yacht and winner of the prestigious AAA 5 Diamond award. Beamed ceilings and an inviting and gracious décor provide a wonderful setting to enjoy savoury Prime Steaks to please the most discerning steak connoisseur. Service is impeccable, but friendly. The beautiful outdoor terrace and adjoining gardens is an ideal spot for a pre-dinner cocktail.

Perched above the beach on the South Shore, Ocean Club is an enticing twist of contemporary Seafood Fusion style. Relax in chic décor surrounded by an inspiring ocean view. Savour tantalising delights that blend Asian ingredients and European techniques, giving each bite an international explosion of flavours.

1. Beach view
2. Waterlot inn steakhouse
3. Resort plan

1. 海滩
2. 水上牛排餐厅
3. 度假村平面图

1. Jasmine Lounge
2. Conference room
3. Ballroom
4. Bermudiana Board Room
5. Newport Room Restaurant
6. Sound Restaurant
7. Administrative offices

1. 茉莉休息室
2. 会议室
3. 宴会厅
4. 贝尔木戴安娜会议室
5. 新港客房餐厅
6. Sound餐厅
7. 行政办公室

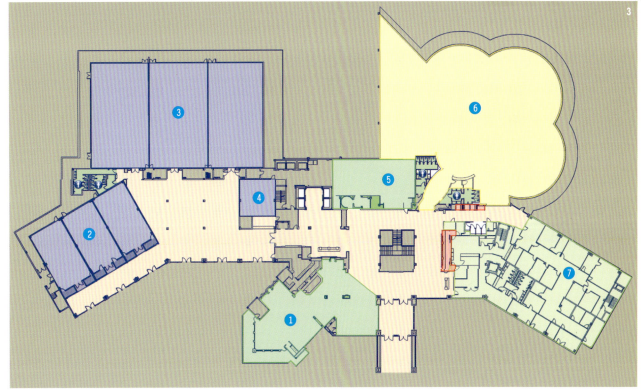

1. Lobby
2. Lobby
3. Ballroom

1. 大堂
2. 大堂
3. 宴会厅

南安普顿费尔蒙特度假村的设计保守，以罕见而深受欢迎的华丽风格为特色，辅以丰富的室内陈设和宽敞的空间感。宏大的楼梯连接着三层楼的公共大厅。奢华的客房设在大楼的两侧，围绕着中央核呈辐射状展开。这一设计让每套客房都享有能够纵观水景的私人游廊。客房十分宽敞，配有超乎想象的大阳台和一张超大的双人床（或是两张普通双人床）。费尔蒙特金楼是酒店中的酒店，为顶级宾客在酒店顶楼提供极致的奢华体验，小到鞋油、大到私人入住手续，无一不做到极致的贴心。金楼提供免费的欧式早餐、报纸和传真、录影带使用权。

带有一到两间卧室的百慕大套房提供独立的客厅区域，配有两台电视和一个小冰箱。柔软舒适的沙发、座椅和原创海洋艺术品让套房显得十分豪华。这些百慕大套房以两个私人阳台和两间设备齐全的大理石浴室、步入式衣柜、书桌、两条电话线（ 象用于高速上网，一条为便利的无绳电话）为特色。各个楼层都有百慕大套房。

优雅的新港餐厅以豪华游艇为蓝本进行设计，是著名的AAA5钻得主，提供浪漫的晚宴。横梁支撑的天花板和富有魅力的装饰为挑剔的美食家享用可口的顶级牛排提供了完美的背景。这里的服务无懈可击而又热情洋溢。美丽的露天平台和紧邻的花园是享用餐前酒的绝佳地点。

海洋俱乐部栖息在南海岸的沙滩上，是融合了现代海鲜餐厅风格的迷人之地，让人们在别致的装饰与灵动的海景中得到充分的放松。结合了亚洲食材和欧式技术的美食让人的感官得到无尽的满足。

1. Restaurant
2. Restaurant
3. Restaurant
4. Restaurant
5. Restaurant

1. 餐厅
2. 餐厅
3. 餐厅
4. 餐厅
5. 餐厅

Grand Velas Riviera Maya
瑞维拉格兰德维拉斯度假村

Completion date: 2008
Location: Riviera Maya, Mexico
Designer: ELIASELIAS AR
Area: 837,699 sqm

竣工时间：2008年
项目地点：墨西哥，瑞维拉
设计师：埃利亚斯利亚斯建筑事务所
项目面积：837,699平方米

Set on over the protected mangroves, jungle and natural freshwater wells locally known as cenotes, Grand Velas enjoys over 300 metres of pristine white sand beach. Diverse luxury accommodation options include uniquely exotic garden suites, ocean-view suites around the central infinity pool, and oceanfront suites for guests age 12 and up with private plunge pools overlooking the azure Mexican Caribbean Sea.

There are a total of 508 suites, ranging in size from 110 to 350 square metres. The suites are divided in three sections within the resort: 1. The Grand Class (adult-only hotel) with 92 oceanfront suites; 2. The Ambassador Hotel,

with 200 oceanfront suites; 3. The Master Hotel, with 216 suites in the jungle with views to a vibrant jungle vegetation and ponds. In all of the three sections, guests enjoy private terraces with plunge pools and incredible ocean or jungle views; space for in-room massage tables; whirlpool bathtubs, and remote-controlled window coverings.

The resort offers extensive amenities: a 7,120-square-metre spa with 40 treatment rooms and suites, men's and women's hydrotherapy facilities, dry saunas, cold plunge pools, vitality pools, whirlpools, herbal steam rooms with colour-therapy, clay and ice rooms and experience showers. There are three life fitness gyms with state-of-the-art equipment. It has a convention centre with 8,496 square metres of flexible spaces for indoor and outdoor meetings and events from 10 to 3,000 guests.

By way of its 12 food and beverage options, the resort offers a variety of dining experiences in upscale restaurants: Gourmet French Cuisine, French Brasserie, Signature/Author Restaurant, Mexican Fine Dining, Asian/Thai/Japanese Dining, Italian Cuisine, plus Lobby Bars, Karaoke Bar, and open-air snack bars around the property all in a contemporary and relaxing atmosphere for every mood.

1. Aerial view of the resort
2. Resort plan

1. 度假村鸟瞰图
2. 度假村平面图

1. Main entrance	1. 主入口
2. Lobby	2. 大堂
3. Lobby & piano bar	3. 大堂和钢琴吧
4. Suites	4. 套房
5. Pool	5. 游泳池
6. Restaurant	6. 餐厅
7. Snack	7. 甜点吧
8. Gym	8. 健身房
9. Teens club	9. 青少年俱乐部
10. Kids club	10. 儿童俱乐部
11. Lifts	11. 电梯
12. Back of the house	12. 后台
13. Beach	13. 海滩

格兰德维拉斯度假村在红树林、热带丛林和天然淡水井（当地称为"天然井"）的环境之中，享有300余米的原始白色沙滩。多种多样的奢华住宿：异域花园套房、海景套房（环绕着中央无边界泳池）和海滨套房（为12岁以上宾客准备，设有远眺墨西哥加勒比海的私人游泳池）。

度假村共有508套客房，大小从110平方米到350平方米不一。客房分布在度假村的三个区域中：1. 设有92套海滨套房的格兰德顶级酒店（仅供成年人入住）；2. 设有200套海滨套房的大使酒店；3. 设有216套热带丛林套房的主人酒店。在三个客房区域中，宾客都能享受带有游泳池的私人露台和非凡的海景或丛林景色、室内按摩台、漩涡浴缸和遥控窗帘。

度假村提供了丰富多彩的娱乐设施：7,120平方米的水疗中心设有40间治疗室和套房、男士和女士水疗设施、汗蒸、冷水池、活力池、漩涡水池、提供色彩疗法的草药蒸汽浴室、粘土室和冰室以及体验淋浴。度假村拥有三间配备着最新装置的健身房。会议中心8,496平方米的灵活空间适用于各种室内外会议活动，可容纳宾客3,000人。

度假村的高档餐厅为宾客们提供了12种餐饮选择：法式美食餐厅、法式啤酒屋、署名/作者餐厅、墨西哥精品餐厅、亚洲/泰国/日本餐厅、意大利美食餐厅、大堂吧、卡拉OK吧和露天甜点吧。这些餐饮设施为宾客们营造了现代而放松的氛围。

1. Pool
2. Pool and rooms view
3. Pond

1. 游泳池
2. 游泳池和客房
3. 池塘

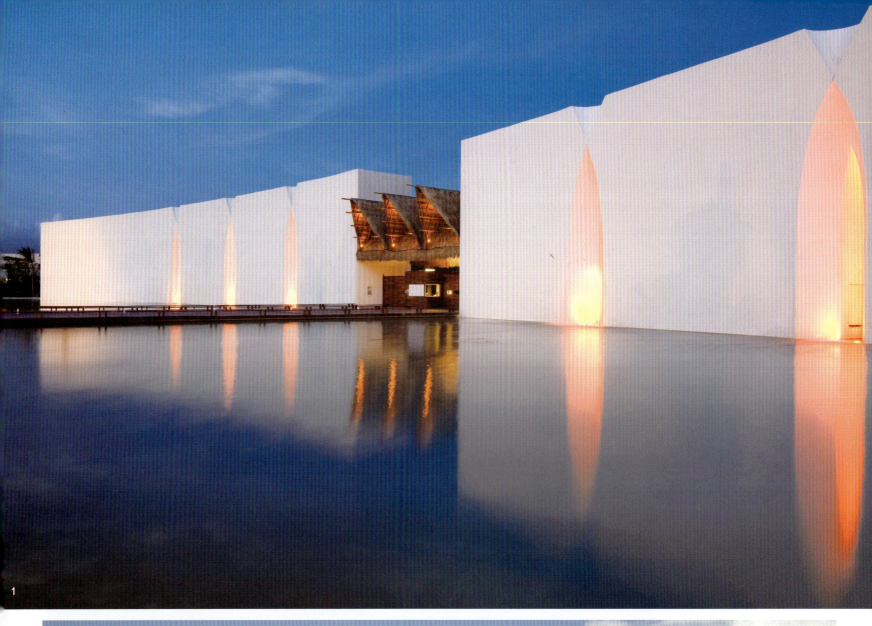

3

1. Resort entrance
2. Suite balcony
3. Spa sensations path

1. 度假村入口
2. 套房阳台
3. 水疗馆感官小路

3

1. Lobby
2. Suite bedroom
3. Signature/Author Restaurant

1. 大堂
2. 套房卧室
3. 署名/作者餐厅

Capella Pedregal de Cabo San Lucas

卡波圣卢卡斯嘉佩乐佩德雷加尔度假村

Completion date: 2009
Location: Cabo San Lucas, Mexico
Designer: HKS, Inc.
Photographer: Robert Reck, Capella Pedregal
de Cabo San Lucas
Area: 18,211 sqm

竣工时间：2009年
项目地点：墨西哥，卡波圣卢卡斯
设计师：HKS公司
摄影师：罗伯特·莱克，卡波圣卢卡斯嘉佩乐佩德
雷加尔度假村
项目面积：18,211平方米

Nestled on the Baja peninsula between the Pacific Ocean and the Sea of Cortez, the ocean provides a backdrop for Capella Pedregal resort in Cabo San Lucas, Mexico, built into the rugged and rocky mountains.

Designed by HKS Hill Glazier Studio, the resort combines a 66-suite hotel, 31 fractional-ownership units and 20 whole-ownership residences into an upscale, luxurious experience for visitors and residents alike.

The all-suite hotel offers all guests ocean views and a private or plunge pool. Three superior waterfront units as well as three beachfront bungalows are offered. Guests are treated to the ultimate in style and relaxation with a

mountaintop spa, state-of-the-art fitness centre and salon. A resort pool and sun deck link the hotel and the private residence club while the resort restaurant features an impressive view of the pool and ocean.

A modern interpretation of Pacific coastal Mexican vernacular architecture using indigenous materials, the resort blends with the ruggedness of the site, creating minimal visual impact and harmonising with the environment.

All the materials used are from the region and crafted by local artisans – the only way to create a true, authentic experience. The beautiful gold's and ochre's of the local quarries, from which the stone for the buildings were extracted, inspired both the exterior and interior colour scheme. Local craftsmen provided everything from sculptures to hand-carved doors, allowing the guest to become immersed in the local culture.

The mountains presented an unusual challenge – the steep elevation change was not passable. To meet this obstacle and preserve the existing landscape, a passageway was carved through the mountain to reach the oceanfront beyond. Hotel guests drive through the passage to access the resort. Exiting the tunnel, guests are greeted with a stunning ocean view framed by two residence buildings.

The site is comprised of a series of small buildings that help to not only preserve the rocky surroundings, but sit better in the steep terrain. This creates a unique setting for each building and achieves incredible, unobstructed views to the Pacific Ocean for them all.

1

1. Resort plan
2. Beach

1. 度假村平面图
2. 海滩

嘉佩乐佩德雷加尔度假村坐落在太平洋和科特斯海之间的巴哈半岛，在起伏的高山掩映之中，海洋为度假村提供了完美的背景。

度假村由HKS希尔·格雷泽尔工作室设计，汇集了一座66套房的酒店、31个分属独立单元和20座全属独立的别墅，为游客和定居者提供了高级奢华体验。

套房酒店为宾客提供了海景和私人游泳池，以及三个高级滨水单元和三座海滨别墅。山顶温泉水疗中心、最先进的健身中心和沙龙为宾客提供了极致的时尚休闲体验。度假泳池和日光浴平台将酒店与私人别墅俱乐部连接起来，而度假餐厅则享有泳池和海景的绝妙景色。

度假村对墨西哥太平洋沿岸的本土建筑进行了现代改造，利用当地材料，与当地的粗犷特色结合起来，将视觉效果最小化，与环境和谐一致。

度假村所采用的全部材料都来自当地，并由当地工匠亲手打造，确保了真实的体验。当地石材漂亮的金色和赭石色为建筑的室内外设计提供了色彩主题。当地工匠一手包办了从雕塑到手工雕刻门的所有元素，让宾客沉浸于当地文化之中。

山脉为设计提供了前所未有的挑战——陡峭的地势难以超越。为了解决这一障碍并保持原有的景观，设计师在山中开凿了一条直达海边的通道。酒店宾客通过这条通道到达度假村。走出通道，宾客便能在两座大楼之间看到迷人的海景。

度假村由一系列小型建筑组成，既保留了山脉的景色，又更适合陡峭的地势。这一设计为每座建筑都打造了独特的背景，保证了每座建筑都能享有太平洋上梦幻的景色。

1. Resort pool
2. Freestanding Residence plan 3BR – level 1
3. Motor court
4. Lobby water feature

1. 度假村游泳池
2. 独立住宅3BR一层平面图
3. 停车大门
4. 大堂水景

1. Entrance	1. 入口
2. Kitchen	2. 厨房
3. Water closet	3. 洗手间
4. Washing room	4. 洗涤室
5. Bathroom & water closet	5. 浴室和洗手间
6. Guestroom	6. 客房
7. Living room	7. 客厅
8. Dining room	8. 餐厅
9. Terrace	9. 露台
10. Swimming pool	10. 游泳池
11. Rest area	11. 休息区

1. Pool bar
2. Freestanding Residence plan 4BR – level 1
3. Residence exterior
4. Estrella Suites stairs
5. Suite terrace

1. 泳池吧
2. 独立住宅4BR一层平面图
3. 住宅外观
4. 爱丝特雷娜套房楼梯
5. 套房露台

1. Stairs
2. Bathroom
3. Single bedroom
4. Terrace
5. Swimming pool
6. Living room
7. Washing room
8. Double bedroom

1. 楼梯
2. 浴室
3. 单人卧室
4. 露台
5. 游泳池
6. 起居室
7. 洗涤室
8. 双人卧室

1. Reception
2. Ocean view
3. Ocean view
4. Spa Sauna tub

1. 前台接待处
2. 海景
3. 海景
4. 水疗桑拿浴室

1. Don Manuels Lounge
2. Beach Club Restaurant
3. Three Meal Restaurant dining room
4. Don Manuels Wine Room

1. 曼纽尔先生休息室
2. 海滨俱乐部餐厅
3. 三餐餐厅
4. 曼纽尔先生品酒室

1. Presidential Suite
2. Residence living room
3. Residence living room
4. Freestanding Residence plan – level 2

1. 总统套房
2. 住宅客厅
3. 住宅客厅
4. 独立住宅二层平面图

1. Stairs
2. Bathroom
3. Single bedroom
4. Terrace
5. Swimming pool
6. Washing room
7. Rest area
8. Double bedroom

1. 楼梯
2. 浴室
3. 单人卧室
4. 露台
5. 游泳池
6. 洗涤室
7. 休息区
8. 双人卧室

1. Ocean-view bedroom
2. Ocean-view bedroom
3. Estrella Suite
4. Presidential Suite master bathroom
5. Suite bathtub

1. 海景卧室
2. 海景卧室
3. 爱丝特雷娜套房
4. 总统套房主浴室
5. 套房浴缸

Esperanza, an Auberg Resort

埃斯佩兰萨度假村

Completion date: 2008
Location: Los Cabos, Mexico
Designer: Wilson Associates
Photographer: Michael Wilson Photography

竣工时间：2008年
项目地点：墨西哥，洛斯卡波斯
设计师：威尔森建筑事务所
摄影师：迈克尔·威尔森摄影

One of the most highly acclaimed resort properties in Los Cabos, Mexico, the Esperanza Auberge caters to a discerning clientele that expects the ultimate in style, elegance and luxury. Wilson Associates approached the project with the goal to create a spectacular space, reminiscent of a Mexican seaside estate, which also embraced and complimented the natural surroundings along the Sea of Cortez.

The interiors were designed to remain consistent with the existing architecture. The Esperanza's generously proportioned windows and openings bring the outside in, allowing for expansive ocean views from each room.

The colour palette remained fresh, revealing nature as art. Indigenous materials were used throughout the interiors, including the use of inlay pebble work, stucco, wood palapa-style ceilings and plaster work.

Each casita features a built-in king bed and night stands, as well as a two-person daybed (chaise) situated in front of the bed. Private outdoor dining is offered on the balcony, which also includes built-in banquettes as well as hammocks for lounging. The casita guestrooms display tremendous attention to furnishings and finishes. The luxurious bathrooms feature large open showers crafted of stone, each with dual showerheads and large windows that look out into the beach.

Another highlight of each room is the use of specially selected artwork and sculpture created by local artists. Guests may purchase any of these art pieces through the Esperanza art gallery located on the property. Local artists help staff the gallery to discuss their art and answer guests' questions.

Each of the new one-bedroom casita suites consists of a bedroom, bathroom, walk-in closet, desk workspace, and an outdoor terrace offering private dining and a hot tub. The views from each suite are spectacular, each with its own unobstructed ocean view.

A highlight of the new construction at Esperanza is the penthouse suite, which boasts phenomenal panoramic ocean views with the ultimate in privacy. The penthouse bedrooms include an iron canopy bed with beautiful white sheers hanging along each side to create a more intimate residential feeling. Beautiful hand-crafted furnishings are placed throughout this space, and large, walk-in closets with built-in drawers and shelving to offer a more welcome and at-home feeling. Guests access the space via their own private lift.

1. Beach view
2. Fountain area

1. 海滩
2. 喷泉

作为墨西哥洛斯卡波斯最富盛名的度假村之一，埃斯佩兰萨度假村为眼光敏锐的宾客提供了时尚、优雅而奢华的终极体验。威尔森建筑事务所以打造非凡的空间为目标，打造了具有墨西哥风情的海滨建筑。度假村与科特斯海边的自然环境紧紧拥抱在一起。

室内设计与原有的建筑风格一致。埃斯佩兰萨宽大的窗口让户外风景融入室内，令每间房间都能够享有丰富的海景。设计的色调新鲜，将自然展现为艺术。本地材料的使用贯穿整个室内设计，其中包括镶嵌卵石、灰泥、木质草棚风格天花板和石膏。

每座别墅小屋都以其嵌入式大床和床头柜为特色，同时床前还摆放着双人躺椅。私人露天就餐区被设在阳台，阳台里还设有嵌入式长凳和休息吊床。别墅客房的设计十分注重室内陈设和装饰。奢华的浴室以巨大的石淋浴间为特色，淋浴间配有双人水龙头，巨大的窗口可以望向海滩。

房间内的另一个重点在于精选的本地艺术家所设计的艺术品和雕塑。如果喜欢，宾客可以向埃斯佩兰萨画廊购买这些艺术品。当地艺术家将在画廊里讨论自己的艺术品并回答宾客的问题。

全新的一室别墅套房由卧室、浴室、步入式衣柜、书桌区、露天平台（提供私人就餐）和热水浴缸组成。每间套房都享有无与伦比的海洋美景。

埃斯佩兰萨的全新特色体现在顶层套房。顶层套房以壮观的海洋全景和极致的私密感而著称。套房卧室里，遮篷铁床上铺着洁白的床单，营造出更加亲近的居家感。美丽的手工陈设品遍布整个空间；巨大的步入式衣柜带有抽屉和衣架，进一步打造了热情而舒适的氛围。宾客们通过私人电梯进入这一空间。

1. Outdoor lobby reception
2. Cacina del Mar Restaurant
3. Fire pit
4. Penthouse floor plan

1. 户外大堂接待处
2. 卡西纳德马餐厅
3. 篝火台
4. 顶楼平面图

1. Terrace
2. Bathroom
3. Main room
4. Closet
5. Lift
6. Spa
7. Bar
8. Lounge

1. 露台
2. 浴室
3. 主卧
4. 橱柜
5. 电梯
6. 水疗中心
7. 吧台
8. 休息室

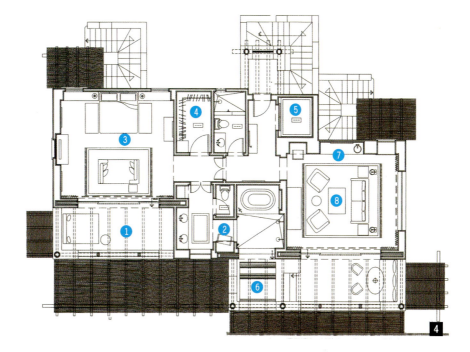

1. Penthouse living room
2. Standard guestroom
3. Penthouse bedroom

1. 顶楼套房客厅
2. 标准客房
3. 顶楼套房卧室

Casa de Campo

田园之家度假村

Completion date: 2009
Location: La Romana, Dominican Republic
Designer: Oscar de la Renta
Photographer: LHW
Area: 28,000,000 sqm

竣工时间：2009年
项目地点：多米尼加，罗马纳
设计师：奥斯卡德拉伦塔
摄影师：LHW
项目面积：28,000,000平方米

Situated in La Romana on the southeast coast of the Dominican Republic, Casa de Campo (Spanish for "Country House") is a Ponderosa-style, tropical seaside resort.

The hotel has reopened following a major renovation of its rooms and suites, and a fully remodelled lobby. Also unveiled was La Cana Restaurant & Lounge by Il Circo, a fine dining restaurant with both indoor and al fresco dining, managed by the Maccioni family of Le Cirque in New York. This is the Maccioni family's second venture at Casa de Campo, following the beautiful Beach Club by Le Cirque at the resort's Minitas Beach. The fitness

centre has been refurbished and a new Montecristo Cigar Lounge has been added.

The new "elite" rooms feature a modern look, while maintaining the use of rich mahogany woods and Coralina stone. The rooms are equipped with 42' LED flat screen TVs and Blu-ray players and cable. Other features are alarm clock radios with dual dock for iPods and iPhones and high-speed, wireless Internet.

The new main area with its grand open spaces makes the absolute most of the surrounding natural Caribbean beauty in a minimalist design.

The main pool area has been completely renovated to include trellised platforms for early morning and late afternoon yoga classes, and shaded seating areas in natural woods and white canopies and cushions.

Lending gracious tropical Caribbean style, this beautifully appointed Villa residence provides 557 square metres of spacious living with 5 bedrooms and 5-1/2 baths plus full maid's quarters. The open floor plan, Living/Dining Room and grand sitting areas lead to the expansive terrace and pool area. The chef's Kitchen is equipped with mahogany cabinetry and new gas cooking appliances. There are 3 Master Bedroom Suites that open to private terraces and an elegant outdoor garden shower.

They feature king-size beds, walk-in closets, whirlpool jacuzzi, bidet, etc. Additional villa bedrooms are expansive and fully equipped with private bath, dressing room and huge walk-in closets. A loft, separate staff quarters and a 2-car garage complete the residence.

The resort offers a unique variety of different facilities. There are 5 Championship Golf Courses by noted course designer Pete Dye, a World Class Marina and Tennis Facilities, Private Beach, Horse Ranch, Polo Fields, a Championship Level Shooting Centre, Cygalle Healing Spa and many other attractions. Guests could relax in the feeling of being far away from the modern world in a simple, relaxing, yet elegant setting.

1. Aerial view of the resort
2. Aerial view of the resort
3. Resort plan

1. 度假村鸟瞰图
2. 度假村鸟瞰图
3. 度假村平面图

1. Gazebo
2. Pool
3. Jacuzzi
4. Patio
5. Wine bar
6. Master bedroom
7. Lobby
8. Living room
9. Kitchen
10. Laundry
11. Car port

1. 露台
2. 游泳池
3. 按摩浴缸
4. 天井
5. 酒吧
6. 主卧
7. 大堂
8. 客厅
9. 厨房
10. 洗衣房
11. 停车场

1. Arrival entrance
2. Pool Cana view
3. Classic villa

1. 接待入口
2. 迦南游泳池
3. 古典别墅

田园之家度假村坐落在多米尼加共和国东南海岸的罗马纳，是一个北美风格热带海滨度假村。

在一系列客房和大堂翻新之后，酒店重新开张。度假村内的拉加纳餐厅——一家提供室内和露天餐饮服务的精致餐厅也同时开张。这是玛西欧尼家族在田园之家的第二处产业，第一处是米尼塔斯海滩上的海滨俱乐部。健身中心经过重新翻修，并且新增了全新的门特克里斯多雪茄吧。

全新的"精英客房"以其现代的外观为特色，同时又保持使用了大量的红木和克拉丽娜石材。客房内配备有42寸LED平板电视、蓝光DVD播放机和有线电视。闹钟收音机带有苹果播放器和手机的底座，配有高速的无线网络。主要区域宏大的空间能够尽享加勒比海的自然美景。

泳池区经过了整体翻新，新增的交叉平台可供清晨和午后的瑜伽课使用，阴凉的休息区则以其茂密的树木和白色穹顶、靠垫为特色。

美丽非凡的别墅总面积557平方米，共有五间卧室和五个半浴室，配有全套的女仆服务。在开放式布局里，客厅/餐厅和巨大的休息区通往宽敞的平台和游泳池。厨房配有红木橱柜和全新的煤气设备。三间主卧套房都通往私人平台和高雅的露天花园淋浴。它们以双人大床、步入式衣柜、漩涡极可意按摩浴缸和坐浴盆等设施为特色。附加别墅卧室十分宽敞，配有私人浴室、更衣室和巨大的步入式衣柜。此外，别墅内还有独立的员工区和两个车库。

度假村提供各种独一无二的设施，其中包括：五座锦标赛级高尔夫球场（由住摩纳哥球场设计师皮特·戴设计）、世界级码头和网球设施、私人海滩、赛马场、马球场、锦标赛级射击中心、塞嘉利治愈温泉水疗馆等。宾客们可以远离喧嚣的现代世界，在简单、放松而优雅的环境中尽享悠闲的假期。

1. Beach Club Le Cirque
2. Villa Excel service
3. La Cana Lounge
4. Restaurant

1. 圆环海滩俱乐部
2. 卓越的别墅服务
3. 迦南休息室
4. 餐厅

3

4

4

1. Lobby
2. Spa reception
3. Elite Suite
4. Elite Room
5. Elite Room

1. 大堂
2. 水疗中心前台
3. 精英套房
4. 精英房
5. 精英房

Pelican Hill

鹈鹕山度假村

Completion date: 2008
Location: Newport Coast, USA
Designer: Darrell Schmitt
Photographer: LHW
Area: 54,000 sqm

竣工时间：2008年
项目地点：美国，纽波特
设计师：达雷尔·施密特
摄影师：LHW
项目面积：54,000平方米

The architectural style of the Resort is inspired by the works of Andrea Palladio, an Italian architect from the 16th century. The beauty and harmony you feel is the result of thousands of small details – from the Italian plaster finish that enhances with age to the handmade terra cotta finials sculpted by a master ceramicist.

With views of the ocean or the Italian-inspired landscape and all the comforts of home, the allure of the Bungalow Guestroom is clear. With residential touches such as a fireplace and soaring 15' wood-beam ceilings, it feels relaxed and luxurious, all at the same time. A spacious private terrace invites you to savour the quiet morning hours or the glorious sunset each evening over the Pacific.

The exclusive Villa Clubhouse provides 930 square metres of private luxury for guests staying at the Villas. A large Great Room features plush seating, an elegant limestone fireplace, a comfy reading area and game tables.

No matter the occasion, the service and setting of Pelican Hill® combine for a memorable business or social affair. Elegant spaces with ocean vistas, private terraces, fireplaces and rustic wood-beam ceilings add a warm residential touch. An event space with a fireplace and rustic wood-beam ceiling sets the tone of warmth and intimacy. A flexible space that is adjacent to Mar Vista ballroom can be set up a variety of ways

offering an indoor space that captures the beauty of the setting through large view-framing windows. The function space in the main building skillfully echoes the comforting residential style of the Resort with warm colours, rich patterned rugs, wood and iron accents. Elegant spaces include a flexible ballroom, three meeting rooms and two boardrooms that spill out onto light-filled loggias featuring floor-to-ceiling windows that overlook manicured greens and views of the Pacific. It is a matchless setting for unforgettable social and business gatherings.

1

1. Resort entry	7. Camp pelican	13. Café	1. 度假村入口
2. Pelican grill	8. Coliseum pool & grill	14. Concierge	2. 鹈鹕烤肉店
3. Café II	9. Andrea restaurant & bar	15. Newsstand & gift shop	3. 二号咖啡厅
4. Golf clubhouse	10. Allegra	16. La cappella	4. 高尔夫俱乐部
5. Golf parking	11. The spa & fitness centre	17. Mar Vista	5. 高尔夫停车场
6. Bungalow guestrooms & suites	12. Lobby & Great Room		6. 别墅客房和套房

7. 鹈鹕露营区	13. 咖啡厅	
8. 竞技场游泳池和烧烤区	14. 门房	
9. 安德里亚餐吧	15. 报摊和礼品店	
10. 爱兰歌娜	16. 小礼拜堂	
11. 水疗和健身中心	17. 海景宴会厅	
12. 大堂和休息室		

1. Resort plan
2. Promenade
3. Resort entrance

1. 度假村平面图
2. 散步长廊
3. 度假村入口

鹈鹕山度假村的建筑风格受到了16世纪意大利建筑师安德里亚·帕拉迪奥的启发。成百上千的细节设计凝聚而成了美与和谐——从意大利灰泥装饰到手工制作的赤陶雕塑，无一不经过精雕细琢。

别墅客房极具诱惑，朝向海洋或是意大利风情景观，具有舒适的居家感。壁炉、高耸的0.5米露梁天花板，让客房兼具休闲感和奢华感。宽敞的私人平台邀请宾客尽情享受宁静的清晨或是在午后俯瞰太平洋上的落日。

专属别墅俱乐部提供了930平方米的私人奢华空间。宽敞的休息室以豪华的休息区、优雅的石灰岩壁炉、舒适的阅读区和游戏桌为特色。

无论在何种场合，鹈鹕山的服务和设施都能够为商务社交场合提供难忘的记忆。优雅的空间，辅以海景、私人露台、壁炉和纯朴的露梁天花板，共同营造出家一般的温暖感觉。壁炉和露梁天花板为活动空间奠定了温暖而私密的氛围。紧邻海景宴会厅的灵活空间通过巨大的窗口为室内空间引入了美丽的风景。主建筑的功能区巧妙地与度假村的家居风格遥相呼应，拥有温暖的色彩、图案丰富的地毯，点缀着木制品和铁制品。度假村设有灵活的宴会厅、三间会议室和两间大型会议室，一直延伸到光亮的凉廊。独特的落地窗俯瞰着修剪整齐的绿化区和太平洋的风景。这是一个无与伦比的社交和商务活动场所。

1. Villa club
2. Villa exterior
3. Lobby lounge
4. Andrea dining terrace

1. 别墅俱乐部
2. 别墅外观
2. 大堂休息室
3. 安德里亚就餐平台

1. Laguna conference room
2. Café
3. Spa women colonnade
4. Couple treatment room

1. 拉古那会议室
2. 咖啡厅
3. 水疗中心女士区廊柱
4. 双人理疗室

1. Villa living room
2. Bungalow bedroom
3. Villa bedroom

1. 别墅客厅
2. 海滨小屋的卧室
3. 别墅卧室

The Kahala Hotel & Resort

凯海兰酒店度假村

Completion date: 2009
Location: Hawaii, USA
Designer: WATG
Photographer: The Kahala Hotel & Resort
Area: 17,280 sqm

竣工时间：2009年
项目地点：美国，夏威夷
设计师：WATG
摄影师：凯海兰酒店度假村
项目面积：17,280平方米

The Kahala Hotel & Resort rests at the very end of one of Oahu's most scenic drives around the iconic Diamond Head Crater landmark. The resort is situated among exotic gardens, complete with a waterfall and lagoon. It completed a $52-million renovation project on its rooms, signature suites, and public areas in 2009.

There are five restaurants and lounges including the award-winning Hoku's, The Veranda lounge, the poolside Seaside Grill and the ocean-front Plumeria Beach House. The resort also features the world-class Kahala Spa with ten private spa suites, and the beachside CHI Health Energy Fitness Centre.

The resort's "Kahala chic" room design features clean,

classic interiors with a Hawaiian touch, and reflects the elegance island lifestyle so richly enjoyed in residential Kahala. The Kahala's Tower and Dolphin wings feature 338 elegant rooms with contemporary comforts. The room décor features ivory and coffee colour tones with hibiscus-patterned wall-to-wall carpeting in the guestrooms and Brazilian walnut (Ipe) wood flooring as well as hand-woven dhurrie rugs with tropical flower patterns in the suites.

The Kahala Kai and Kahala Beach Suites offer increased seclusion and privacy near the Dolphin Lagoon and are only steps away from beautiful Kahala Beach. The Kahala Beach Suite's dark wood and earth tone fabrics with anchor pieces designed by Nicole Miller create a contemporary and distinctly island-style ambience complete with dining table and an outdoor barbecue grill perfect for entertaining family. On the second level offering stunning ocean views, the Kahala Kai Suite, with its cosy beach cottage décor by Ralph Loren Polo Collection, is chic and trendy, reflecting relaxed coastal living at its very best.

Boasting a magnificent ocean view is the Presidential Suite, which gives a residential grand beach estate feel. The two-bedroom space features dark woods, clean white details and a maritime colour palette. In addition to its spacious layout and top-of-the-line amenities, special

1. Beach view
2. Partial Ocean View Lanai plan
3. Ballroom
4. Ballroom

1. 海景
2. 海景套房局部平面图
3. 宴会厅
4. 宴会厅

touches make it a Kahala Signature Suite: rainforest leaf-patterned Nepalese pile carpet, a set of original-bound Winston Churchill volumes, seashell-inspired glass ornaments, fully stocked bar with top-shelf liquor, and a library of photographs and books on the Hawaiian Islands.

The Imperial Suite, where both Presidents and Princesses have slept, combines luxury sleeping accommodations with amenities for business and private entertaining. The Imperial Suite is a spacious, 200-square-metre sanctuary offering separate master bedroom, living room, and office spaces designed with Hawaiian residential sensibilities for home-away-from-home comfort. The apartment features commanding views of the azure-blue Pacific Ocean from Koko Head to Diamond Crater.

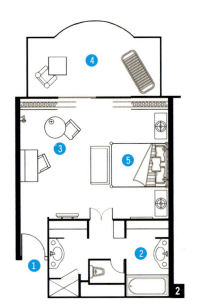

1. Entrance
2. Bathroom
3. Living room
4. Terrace
5. Bedroom

1. 入口
2. 浴室
3. 客厅
4. 露台
5. 卧室

凯海兰酒店度假村位于奥阿胡斯环绕着钻石头火山口的景观路的尽头。度假村四周环绕着异域花园、瀑布和潟湖。2009年,度假村花费了5200万美元对客房、特色套房和公共区域进行了翻新。

度假村中有五家餐厅和休闲吧,其中包括凯海兰高级星餐厅、凉台酒吧、海滨烧烤餐厅以及知子花爱心餐厅。顶级的凯海兰水疗中心设有10个水疗套房以及伊尔滨海CHI健康动力健身中心。

每套客房和套房都浓缩了度假村的经典"凯海兰时尚"设计——兼具现代外观和开放的热带风情,反映了凯海兰优雅的岛屿生活方式。凯海兰塔和海豚套房拥有338间优雅的客房,配有全套现代设施。客房装修以象牙色和咖啡色为主。客房采用了木槿花图案的满铺地毯;套房则采用了巴西胡桃木地板和带有热带花纹的手纺纱棉毯。

凯海兰卡伊和凯海兰滨海套房在海豚湾附近提供了私密而舒适的隐居之所,距离美丽的凯海兰海滩仅几步之遥。凯海兰滨海套房的深色木纹和大地色系织物配合着由尼克尔·米勒设计的装饰品,营造出现代而具有小岛风情的氛围。餐桌和露天烧烤台将为家庭提供更多的欢愉。凯海兰卡伊套房的二楼空间提供了绝妙的海洋美景,内部装饰以舒适的海滨小屋风格为主,拉尔夫·洛伦·保罗的装饰品时尚而别致,凸显了轻松的海岸居住氛围。

总统套房的壮丽海景令人惊叹,拥有滨海豪宅的感觉。两间卧室以深色木家具、简洁的白色细部设计以及海洋色调为主。除了宽敞的布局和顶级的设施之外,套房还拥有显著的凯海兰特色:带有雨林叶子图案的尼泊尔容貌地毯、限量版温斯顿·丘吉尔全集、贝壳玻璃装饰品、摆满好酒的吧台以及以夏威夷图片和图书为主的图书室。

皇室套房曾是总统和公主们的下榻之处,里面结合了奢华的卧室设施和商务私人娱乐设施。皇室套房十分宽敞,200多平方米的空间内设有独立主卧室、客厅和办公空间。满溢着夏威夷风情的套房提供了家一般的舒适感。套房拥有从可可山到钻石头的蔚蓝的太平洋美景。

1. Lobby and lounge
2. Restaurant
3. Ko'olau Partial Ocean View Suite plan

1. 大堂和休息室
2. 餐厅
3. 库劳海景套房平面图

1. Entrance
2. Bathroom
3. Living room
4. Terrace
5. Bedroom
1. 入口
2. 浴室
3. 客厅
4. 露台
5. 卧室

1. One-bed guestroom
2. Living room
3. Two-bed guestroom
4. Koko Head Ocean-Front Suite plan
5. Koko Head Ocean-Front plan

1. 单人客房
2. 客厅
3. 双人客房
4. 可可山海景套房平面图
5. 可可山海景房平面图

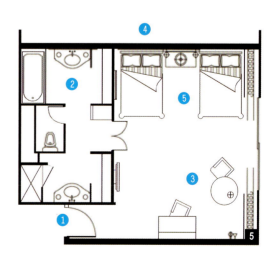

1. Entrance	4. Terrace	1. 入口	4. 露台
2. Bathroom	5. Bedroom	2. 浴室	5. 卧室
3. Living room		3. 客厅	

1. Entrance	4. Terrace	1. 入口	4. 露台
2. Bathroom	5. Bedroom	2. 浴室	5. 卧室
3. Living room		3. 客厅	

The Betsy South Beach

贝琪南海滩度假村

Completion date: 2009
Location: Miami Beach, USA
Designer: Diamante Pedersoli and Carmelina Santoro
Photographer: Moris Moreno
Area: 4,200 sqm

竣工时间：2009年
项目地点：美国，迈阿密海滩
设计师：戴亚曼特·比德尔索利和卡尔米丽娜·桑托罗
摄影师：莫里森·莫雷诺
项目面积：4,200平方米

One of The Betsy's many assets are the enormous floor-to-ceiling windows which stretch across the expansive façade allowing natural light to flow through the space. The tropical colonial design of the lobby is obtained by the use of wooden plantation-style shutters as dividing walls which separate the hotel lobby and guest reception from BLT Steak. Lighting consists of Ivory-coloured Fortuny Sheherazade chandeliers hanging from the coffered ceilings and these elegant fixtures are complemented by colonial-style fans in the lobby lounge. The historical terrazzo flooring is accented with natural fibre area rugs. Wooden antique world globes, zebra framed mirrors, safari pictures and potted palms complete

the theme.

The Betsy's 61 rooms and suites are private, beachside havens. Behind white wooden plantation shuttered windows, the contemporary colonial design wraps every guest in exquisite detail. Every room features dark black walnut hardwood floors, white lacquered dressers, cabinetry and desks, and stately poster beds. Four different colour schemes are used to define the clean white rooms: fields of lavender in Provence, sunbathed Tuscan ochre, vibrant sea coral, and fresh Spring green. These inspirations came from brilliant colours in nature. The richly textured linen curtains and pillow shams are bordered in matching colour, while pillows and chairs

of various fabrics further distinguish the colour accents. Raffia-covered ceilings, chairs and headboards lend a unique texture to the rooms. The hotel's accommodations otherwise share the same luxurious amenities and design. Luxury is defined by many factors and the various lighting selections should be highlighted in The Betsy's accommodations. Table and floor lamps, bedside lighting, and high hats (in-ceiling lighting) allow for a personal selection to suit every mood. Fresh chrome and white ceiling fans continue the tropical colonial theme and work alone or in conjunction with air conditioning to carry the breeze from the Atlantic throughout the rooms. Beds are topped with fine Frette® linens, and guests may select

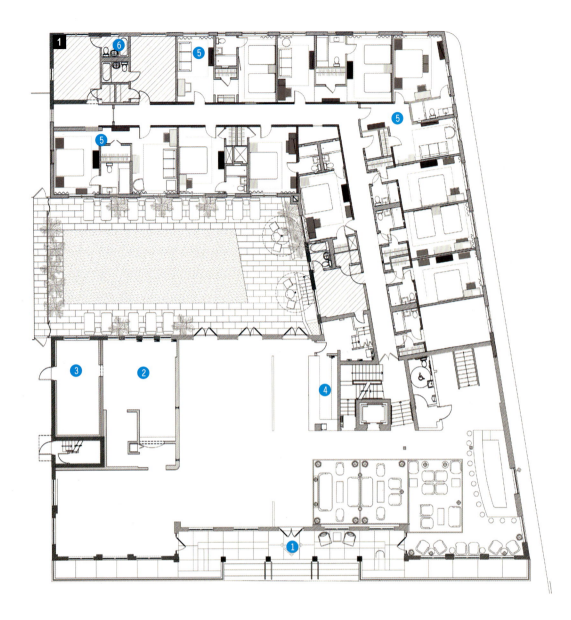

1. Entrance
2. Kitchen
3. Back room
4. Front desk
5. Suite
6. Bathroom

1. 入口
2. 厨房
3. 内室
4. 前台
5. 套房
6. 浴室

from a menu which includes everything from hypoallergenic to down. Plush robes and slippers are standard in each room. Every room and suite at The Betsy combines the technology of today blended seamlessly with the architecture and grace of yesteryear. The resulting product is a fresh concept in South Beach with timeless style.

1. Resort plan
2. Lobby
3. Lobby

1. 度假村平面图
2. 大堂
3. 大堂

贝琪度假村的众多特色之一便是遍布建筑外立面的落地窗，这些窗户为室内空间带来了充足的自然光。大堂的热带殖民地式设计通过作为隔断的农庄风格百叶屏风（将酒店大堂和接待处与BLT牛排餐厅隔开）得以展现。象牙色的吊灯从方格天花板上垂下；大堂休息室里的殖民地风格电扇则进一步凸显了酒店优雅的氛围。古董木地球仪、斑纹框架镜子、狩猎绘画和盆栽棕榈树完善了整个主题。

贝琪的61套客房和套房是私密的海滨天堂。在白色木制百叶窗后面，现代殖民地设计以其精致的细节照料着每位宾客。每间房间都以黑胡桃木硬木地板、白色喷漆梳妆台、家具和书桌、四柱床为特色。简洁的白色房间采用了四种不同的色彩主题：普罗旺斯的薰衣草色、沐浴在阳光中的托斯卡纳土黄色、活力十足的珊瑚礁色和清新的春绿色。这些灵感都源于自然的亮丽色彩。纹理丰富的亚麻窗帘和枕套采用了配套的色彩，而五彩缤纷的枕头和座椅则进一步区分了色调。覆盖着拉菲纤维的天花板、座椅和床头板为房间增添了独特的质感。酒店的客房同样奢华而具有设计感。

奢华体现在许多方面，贝琪度假村的各式灯光设施绝对值得一提。台灯、落地灯、床头灯和天花板嵌灯都可以通过个性设置来配合情绪变化。新鲜的铬黄色和白色吊扇延续了热带殖民地主题，与空调一起为室内带来了大西洋的清风。客房的床上铺着精致的弗雷蒂品牌床单，宾客们也可以根据需要进行自主选择。每间客房内都放置着舒适的长袍和拖鞋。贝琪度假村的每套客房和套房都将当今先进的技术与充满了旧日辉煌的建筑完美地结合在一起。因此，贝琪南海滩度假村的风格清新而又不乏经典。

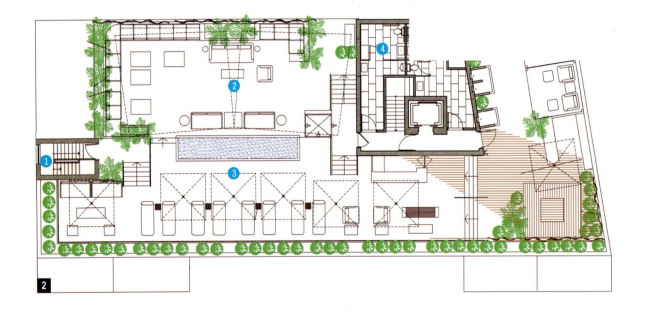

1. Stairs

2. Living room

3. Rest area

4. Washing room

1. 楼梯

2. 休息室

3. 休闲区

4. 卫生间

1. BLT steak restaurant
2. Top floor plan
3. Guestroom
4. Suite plan of second floor

1. BLT牛排餐厅
2. 顶楼平面图
3. 客房
4. 三层套房平面图

1. Suite bedroom
2. Suite living room
3. Suite bathroom

1. 套房卧室
2. 套房起居室
3. 套房浴室

1. Guestroom
2. Living room
3. Guestroom

1. 客房
2. 客厅
3. 客房

Cotton House

纯棉别墅度假村

Completion date: 2010
Location: Mustique Island, Canada
Designer: Chapman Design Group
Photographer: LHW
Area: 52,600 sqm

竣工时间：2010年
项目地点：加拿大，马斯蒂克岛
设计师：查普曼设计集团
摄影师：LHW
项目面积：52,600平方米

The imposing storage house with its foot-thick stone walls and typical veranda became the dining room, the ballroom, and the popular Monkey Bar. In true colonial style, the "rough and ready" was set aside at night, when guests would dress for dinner and be entertained by some sort of show that the irrepressible Tennant had managed to conjure. Attracted by Mustique's promise of privacy, ideal weather and perfect beaches, as well as the advantage of being out of the hurricane zone, many big names were signed on the dotted line. Mick Jagger, David Bowie and Tommy Hager are among those who were approved.

Cotton House is a collection of houses dotted around

a lush estate, each with a view, and guests are accommodated in a house of their own. It has the best location on this man-made Eden, on a promontory that divides windswept L'Ansecoy Bay from the calm green waters on the other side. Surfing and snorkelling are therefore both just a stone's throw away.

Despite the stories of wild times and crazy parties, Mustique's present-day appearance – immaculate and unspoilt – is the result of rigorous efforts. No helicopters are allowed on the island and the landing strip accommodates nothing bigger than a six-seater. The airport, like the nearby church, was built in bamboo,

and there is no traffic to speak of – everyone gets around on a Mule. This is not quite as rustic as it sounds: Mule is the brand name of Kawasaki's motorised golf carts. These only have two gears (forward and back) but there are still speed bumps all over the island – noise is micro-managed like everything else here. Paradise, it seems, takes relentless planning.

Cottages charmingly stand alone West Indian-styled individual rooms with garden views and distant Atlantic Ocean views. Bedrooms feature king-size beds, spacious dressing rooms and each room opens onto a large furnished veranda.

Seafront Rooms are luxury king rooms which overlook the Caribbean Sea. All are furnished with woven rattan signature pieces and feature four-poster beds. Some have plunge pools for added relaxation.

Grenadines Suites are all decorated with French doors opening from both the bedroom and the living room onto a large furnished veranda. The two suites on the lower floor feature private plunge pools which flow from the suites veranda with an elegantly furnished pool deck.

1. Villa pool
2. Resort plan
3. Pool and deck
4. Ocean-front pool deck

1. 别墅游泳池
2. 度假村平面图
3. 游泳池和平台
4. 海景游泳池平台

1. Gazebo	6. Second bedroom	1. 露台	6. 卧室
2. Dining room	7. Bathroom	2. 餐厅	7. 浴室
3. Living room	8. Outdoor shower	3. 客厅	8. 露天淋浴
4. Master bedroom	9. Guest powder room	4. 主卧	9. 客用洗手间
5. Bathroom	10. Butler pantry	5. 浴室	10. 管家备餐室

1. Villa pool
2. Terrace

1. 别墅游泳池
2. 露台

宏伟的栈房搭配着厚厚的石墙和典型的游廊，里面设置着餐厅、宴会厅和深受欢迎的猴子酒吧。度假村沉浸在真正的殖民地风情之中。夜晚，"简陋随意"的风格被抛到一边，宾客们穿上正式的服装享用晚宴并观看魔术表演。受到私密的保证、理想的气候和完美的沙滩的吸引，同时又被远离飓风带而诱惑，许多大人物都愿意来到纯棉别墅度假村，如米克·贾格尔、大卫·鲍伊和汤米·海格等。

纯棉度假村由一系列坐落在茂密的绿植中的别墅组成，宾客可以自由选择独栋别墅。度假村拥有这座人造伊甸园中的最好位置，位于狂风大作的兰斯科伊湾和平静的碧水之间的海角上。因此，冲浪和潜泳是度假村的两大主题。

除了狂欢时刻和派对之外，马斯蒂克岛仍然保留着完美无瑕的纯朴模样。岛上禁止直升机飞行，着陆带最多只能允许六架飞机降落。机场和附近的教堂都由竹子建造而成。岛上除了机动高尔夫球车之外，没有其他交通工具。这种由川崎重工制造的高尔夫球车只有两个传动装置（分别位于前后）。即便如此，岛上还是遍布减速带，以最大限度地减少噪声。马斯蒂克岛在严格的规划下，宛若天堂。

别墅的西印度风格房间极富魅力，享有花园景观和遥远的大西洋海景。卧室以双人大床、宽敞的更衣室为特色，每间房间都通往一个巨大的家居游廊。

奢华的海滨客房俯瞰着加勒比海的美景。客房内采用编织藤条装饰品，以四柱床为特色。一些客房外还设有私人游泳池。

格林纳丁斯套房全部装饰着法式对开门，卧室和客厅由这些门进入巨大的游廊。建筑底层的两座套房以私人游泳池为特色，水流沿游廊流到设计优雅的水池中。

1. Lounge
2. Living room
3. Lounge
4. Lobby

1. 休息室
2. 客厅
3. 休息室
4. 大堂

1. Living room
2. Guestroom
3. Guestroom

1. 客厅
2. 客房
3. 客房

InterContinental Bora Bora
Resort & Thalasso Spa
洲际波拉波拉岛泰拉索度假村

Completion date: 2006
Location: Bora Bora, French Polynesia
Designer: Carole Stévenin and Sylvain Proyart in collaboration
with the architectural firm IIHI
Photographer: Masujima, Danee Hazama, Kenji Kudo
Area: 12,000 sqm

竣工时间：2006年
项目地点：法属波利尼西亚，波拉波拉岛
设计师：卡罗尔·斯蒂文尼；希尔维·普罗亚特；
IIHI建筑公司
摄影师：马苏吉玛；戴尼·哈扎马；工藤健二
项目面积：12,000平方米

This stunning resort is romantically located on Motu Piti Aau (two hearts in Polynesian) on the barrier reef between the ocean and the lagoon. The resort has 80 spacious over-water villas overlooking the azure lagoon. From the "ethno-chic" design of the bungalows to the trendy Bubbles Bar adorned with Philippe Stark furniture, the resort's restaurants, facilities and over-water bungalows are all impeccably designed offering a modern take on Polynesian décor.

Moreover, when International and French cuisine meets up the freshest products from French Polynesia and the sea, it creates a thousand delicious tastes bringing

memorable dining experiences! The resort has something to match the moment and every mood in the two restaurants, two bars and 24-hour room service. A large choice of delicious dishes, not forgetting lighter meals, will satisfy your hunger, with a variety of drinks, exotic cocktails and delicious wines from France and other regions will quench your thirst.

Inspired by Marlon Brando, it is the first hotel in the world producing air-conditioning from water extracted from the ocean. Powered by deep-sea water drawn from a depth of 900 metres, this system saves 90% of the hotel's electricity.

The resort is the only resort in French Polynesia with a private, over-water wedding chapel. The Polynesian wedding ceremonies in the Blue Lagoon Chapel or on the beach are tailor-made especially for the clients. Add candlelit dinners on the beach and private sunset cruises on the lagoon with a bottle of Champagne and signature spa treatments with deep ocean water in an over-water Bungalow for two and you will have the romantic stay perfect.

Designed by Algotherm, the Deep Ocean Spa is the very first Thalasso spa in the world to utilise the benefits of deep-sea water and minerals extracted from the Pacific

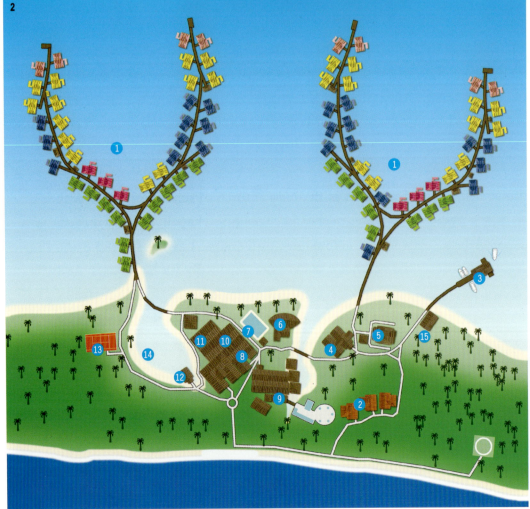

1. Overwater villas   1. 水上别墅

2. Family suites Motu   2. 家庭套房

3. Marina   3. 码头

4. Reception   4. 前台

5. Day room   5. 日间室

6. "Sands" bar & restaurant   6. 沙滩酒吧和餐厅

7. Swimming pool   7. 游泳池

8. "Azur" shop   8. 阿祖尔商店

9. Deep Ocean Spa   9. 深海水疗中心

10. "Bubbles" bar & lounge   10. 泡泡酒吧和休息室

11. Beach lounge   11. 海滨休息室

12. Blue Lagoon Chapel   12. 蓝色珊瑚教堂

13. Tennis   13. 网球场

14. Coral garden   14. 珊瑚花园

15. Bathys centre   15. 洗浴中心

1. Sand's bar
2. Resort plan
3. Dining on the beach

1. 沙滩吧
2. 度假村平面图
3. 沙滩晚餐

Ocean at a depth of more than 900 metres in all of its signature treatments.

The Deep Ocean Spa is the centrepiece of the resort but the resort offers a great many cultural and recreational activities to fill your days including canoeing, kayaking, snorkelling, stargazing and handicraft demonstrations or just relaxing in an amazing location.

这座非凡的度假村坐落在浪漫的默图皮蒂艾尤（波利尼西亚的两座心形岛），在海洋和环礁湖之间的堡礁之上。度假村80座宽敞的水上别墅俯瞰着碧蓝的潟湖。

从别墅民俗加精美的设计风格到布置着菲利普史塔克家具的现代泡泡酒吧，酒店的餐厅、设施和水上别墅都经过了精心设计，在波利尼西亚风情中增添了现代感。

当国际和法式美食与新鲜的法属波利尼西亚产品和海洋相遇，成百上千种美食将为人们带来难以忘怀的餐饮体验。度假村的两家餐厅、两家酒吧和24小时客房服务能够满足任何场合和任何情绪的需求。美味可口的菜品和小食将满足你的味蕾，而各式各样的饮品、具有异国情调的鸡尾酒和来自法国和其他区域的美酒将滋润你干燥的喉咙。

受到马龙·白兰度的启发，这是全球第一家利用海水来进行支持空调系统的酒店。空调以来自900米深海的海水进行驱动，节省了酒店90％的电力。

这是法属波利尼西亚唯一一家带有私人水上婚礼教堂的度假村。在蓝色潟湖教堂或海滩上举行波利尼西亚式婚礼将为宾客量身打造。沙滩烛光晚宴和湖上私人黄昏航行，配上香槟酒和独家双人深海水疗，你将体验到前所未有的浪漫。

深海水疗馆由Algotherm独家设计，是全球第一个使用来自900米深海海水和矿物进行治疗的海水水疗馆。

深海水疗馆是度假村的中心，此外，度假村还提供了许多丰富的文化娱乐活动，例如皮划艇、独木舟、浮浅、观星和手工艺品展览。当然，游客也可以什么都不做，在这里享受完全的放松。

1. Lounge
2. Reception
3. Bubble's bar

1. 休息室
2. 前台
3. 泡泡吧

1. Reef Restaurant
2. Reef Restaurant
3. Reef Restaurant

1. 暗礁餐厅
2. 暗礁餐厅
3. 暗礁餐厅

3

1. Villa lounge
2. Villa bathroom
3. Villa bedroom

1. 别墅休息室
2. 别墅浴室
3. 别墅卧室

InterContinental Moorea Resort and Spa

洲际茉莉雅岛度假村

Completion date: 2010
Location: Moorea, French Polynesia
Designer: Pierre Jean Picart
Photographer: Massujima
Area: 110,000 sqm

竣工时间：2010年
项目地点：法属波利尼西亚，茉莉雅岛
设计师：皮埃尔·珍·皮卡尔特
摄影师：马苏吉玛
项目面积：110,000平方米

Situated on one of the most spectacular tropical islands of French Polynesia, InterContinental Moorea Resort and Spa provides a magnificent setting for a memorable South Seas vacation. The resort features 143 Lanai rooms and garden pool, beach and also over-water bungalows. The over-water bungalows are all impeccably designed offering a modern take on Polynesian décor.

The resort has two restaurants, Fare Nui and Fare Hana, and one bar, Motu Iti Bar. Both restaurants offer French, Tahitian and international cuisine. The resort also offers various Polynesian evenings including the soirée merveilleuse each Saturday, which is a sumptuous

seafood buffet and Polynesian dance show taking place on the main beach, followed by dancing under the stars. Moreover, the resort's Moorea Dolphin Centre offers interactive programmes with bottlenose dolphins, educating both tourists and locals, and is also home to the Sea Turtle Rehabilitation Centre. Sanctioned by the ministry of the environment, this facility serves as a hospital to ill and injured turtles from all the islands in French Polynesia. Additionally, the InterContinental Moorea Resort and Spa recently garnered Silver Green Globe 21 certification an international certification programme for sustainable travel and tourism.

For added maritime fun, the resort also houses Bathy's Diving Moorea. The centre is available for all levels of diving, from novices to experienced divers, and provides all necessary training and equipment.

Unique in its design, the Hélène Spa was created to blend into its lush, tropical surroundings and provide a truly comfortable and relaxing setting for spa guests. The spa is housed in traditional Polynesian fares thatched with palm fronds, and uses natural materials such as precious woods, volcanic stone, and bamboo and offers natural indigenous treatments. This centre brings to the visitors knowledge and a real experience .

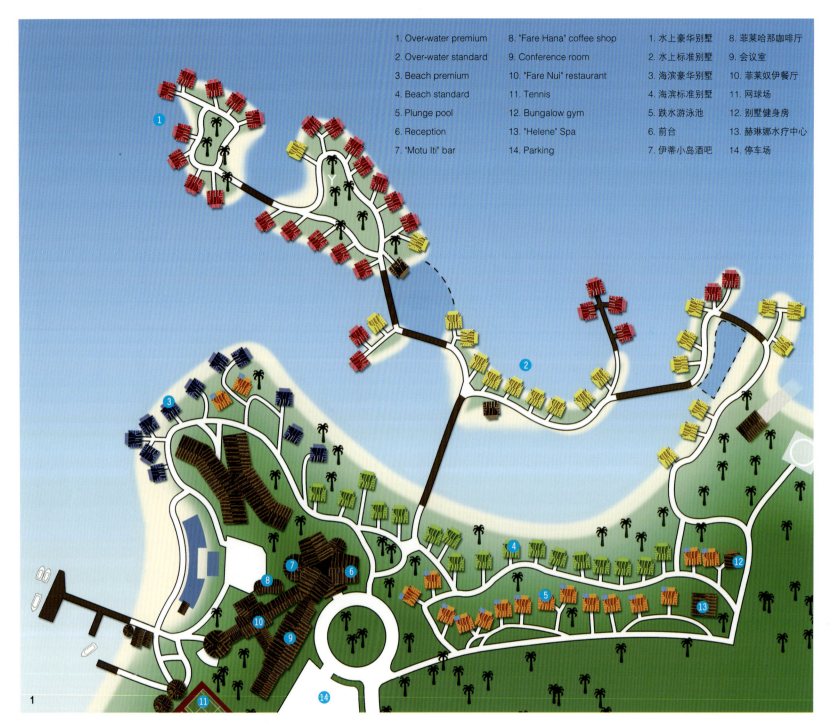

1. Over-water premium
2. Over-water standard
3. Beach premium
4. Beach standard
5. Plunge pool
6. Reception
7. "Motu Iti" bar
8. "Fare Hana" coffee shop
9. Conference room
10. "Fare Nui" restaurant
11. Tennis
12. Bungalow gym
13. "Helene" Spa
14. Parking

1. 水上豪华别墅
2. 水上标准别墅
3. 海滨豪华别墅
4. 海滨标准别墅
5. 跌水游泳池
6. 前台
7. 伊蒂小岛酒吧
8. 菲莱哈那咖啡厅
9. 会议室
10. 菲莱奴伊餐厅
11. 网球场
12. 别墅健身房
13. 赫琳娜水疗中心
14. 停车场

洲际茉莉雅岛度假村坐落在法属波利尼西亚最瑰丽的热带岛屿上，为难以忘怀的南海假期提供了完美的布景。度假村拥有143套蕾娜房，花园泳池别墅，沙滩别墅和水上别墅。水上别墅的设计融合了波利尼西亚风情和现代设计。

度假村拥有两家餐厅——菲莱奴伊和菲莱哈那和一间酒吧——伊蒂小岛酒吧。两家餐厅都提供法式美食、塔希提美食和国际美食。度假村还提供各种各样的波利尼西亚风情之夜，其中周六奇妙之夜将提供豪华的海鲜自助大餐和波利尼西亚舞蹈表演以及星光下的共舞。

此外，度假村的茉莉雅岛海豚中心还将提供与海豚的互动项目，以教育游客和当地居民。海龟复建中心也具有同样的效果。海龟中心获得了环境部的批准，为法属波利尼西亚群岛生病和受伤的海龟提供复建服务。此外，洲际茉莉雅岛度假村还获得了经21银绿色星球认证的可持续旅行和旅游景点认证。

为了增添海洋乐趣，度假村还设有茉莉雅贝琪潜水中心。该中心为新手和老手提供各种层次的潜水，并配有必要的训练和设备。

设计独特的爱琳娜水疗中心融入了郁郁葱葱的热带环境，为宾客们提供了真正的舒适和休闲环境。水疗中心设在由棕榈叶覆盖的传统波利尼西亚凉亭中，利用珍稀木材、火山石和竹子等天然材料为宾客提供当地的特色治疗，既为宾客提供了养生知识，又让他们进行了真正的放松体验。

1. Resort plan
2. Aerial view
3. Bungalow with swimming pool

1. 度假村平面图
2. 鸟瞰图
3. 游泳池别墅

1. Swimming pool bar
2. Fare Hana restaurant
3. Lobby
4. Motu Iti bar

1. 游泳池酒吧
2. 菲莱哈那餐厅
3. 大堂
4. 伊蒂小岛酒吧

1. Lounge and bedroom
2. Bungalow bedroom
3. Bungalow bathroom

1. 客厅和卧室
2. 别墅卧室
3. 别墅浴室

InterContinental Tahiti Resort and Spa

洲际大溪地岛度假村

Completion date: 2005 (renovation)
Location: Tahiti, French Polynesia
Designer: ABC Design Team
Photograph: Massujima
Area: 120,000 sqm

竣工时间：2005年（翻新）
项目地点：法属波利尼西亚，大溪地岛
设计师：ABC设计团队
摄影师：马苏吉玛
项目面积：120,000平方米

InterContinental Tahiti Resort has long been rated the top hotel on the island of Tahiti. It is located alongside the lagoon with a view of Tahiti's sister island, Moorea. The resort features 260 guestrooms, including 16 over-water motu bungalows and 15 over-water lagoon bungalows, including a luxury villa that is a replica of the over-water bungalows at InterContinental Bora Bora Resort & Thalasso Spa.

The resort features two restaurants, Le Lotus and Le Tiare, and three bars, the Tiki Bar, Le Lotus Swim-Up Bar and the Lobby Bar. Le Lotus restaurant is completely secluded at the far end of the resort and features cuisine

from Chef Bruno Schmitt, formerly of three-Michelin star restaurant L'Auberge d'Ill in France. Le Tiare features a performance by Les Grands Ballets de Tahiti, the island's internationally-renowned performance group.

The resort opened the Deep Nature Spa in December 2008. Products and treatments are similar to those of the Deep Ocean Spa by Algotherm located at the InterContinental Bora Bora Resort and Thalsso Spa.

Also, located at the InterContinental Tahiti Resort is the Bathy's Diving Club, the only dive centre in French Polynesia rated "Five-star Gold Palm IDC PADI". This recognition by the world's most important community of divers is partly due to the quality of its establishment, equipment and boats.

The Lagoonarium is unique to the South Pacific and recreates the underwater world found in the ocean by providing a constant flow of seawater filled with corals, shells and tropical fishes. The Lagoonarium houses more than 200 sea species of all sizes and colours including parrot fishes, sting rays and angel fishes. This very delicate ecosystem is regularly controlled and monitored by a team of scientists.

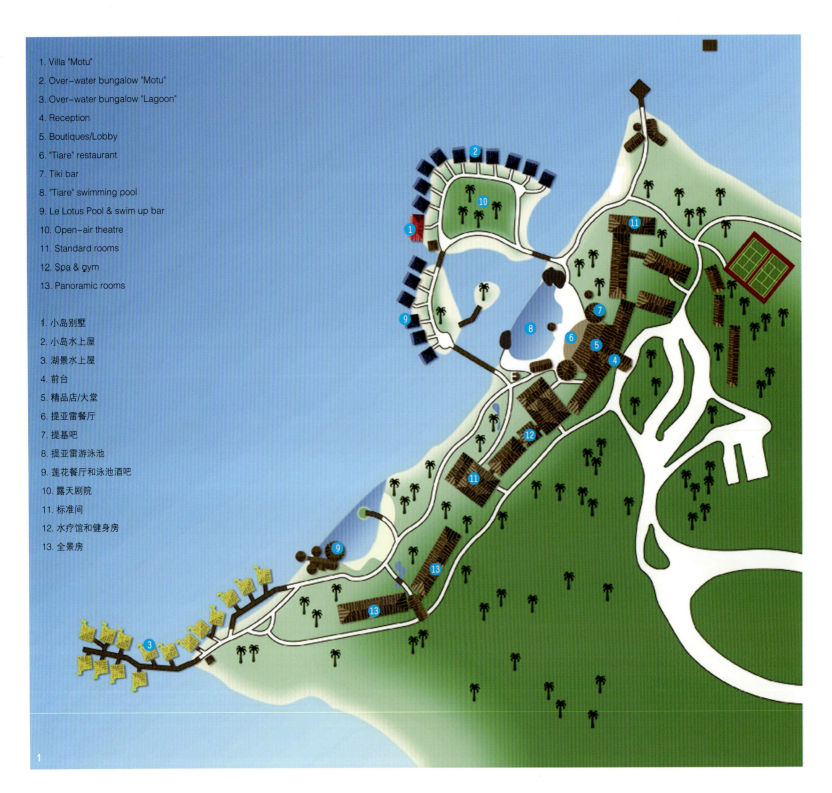

1. Villa "Motu"
2. Over-water bungalow "Motu"
3. Over-water bungalow "Lagoon"
4. Reception
5. Boutiques/Lobby
6. "Tiare" restaurant
7. Tiki bar
8. "Tiare" swimming pool
9. Le Lotus Pool & swim up bar
10. Open-air theatre
11. Standard rooms
12. Spa & gym
13. Panoramic rooms

1. 小岛别墅
2. 小岛水上屋
3. 湖景水上屋
4. 前台
5. 精品店/大堂
6. 提亚雷餐厅
7. 提基吧
8. 提亚雷游泳池
9. 莲花餐厅和泳池酒吧
10. 露天剧院
11. 标准间
12. 水疗馆和健身房
13. 全景房

洲际大溪地岛度假村一直都是大溪地岛的顶级酒店。度假村沿着环礁湖而建，远眺着大溪地岛的姐妹岛——茉莉亚岛。度假村拥有260套客房，其中包括16套水上默图别墅和15套水上湖景别墅。其中的一座奢华别墅完全重现了洲际波拉波拉岛度假村的水上别墅。

度假村有两家餐厅——莲花餐厅和提亚雷餐厅、三间酒吧——提基酒吧、莲花游泳吧和大堂酒吧。莲花餐厅隐蔽在度假村的远角，主厨布鲁诺·施密特（前米其林三星餐厅法国德伊尔餐馆的主厨）将为宾客提供特色美食。提亚雷餐厅以大溪地芭蕾舞团（驰名世界的表演团队）的表演而著称。

度假村在2008年12月开放了深层自然水疗馆，其产品和疗法与洲际波拉波拉岛度假村的深海水疗馆所提供的十分相似。

此外，洲际大溪地岛度假村还设有贝琪潜水俱乐部——法属波利尼西亚唯一的"五星级金棕榈"潜水中心。这个由全球最重要的潜水者协会颁布的认证充分肯定了俱乐部的设施、设备和船舶。

太平洋独特的环礁湖为宾客提供了奇妙的水下体验：流动的海水里满是珊瑚、贝壳和热带鱼。环礁湖内生存着200余种大小、色彩不一的海洋生物，其中包括鹦嘴鱼、魔鬼鱼和天使鱼。一队科学家将定期控制并检测这个细腻的生态系统。

1. Resort plan
2. Tiare restaurant at night
3. Lotus restaurant at night

1. 度假村平面图
2. 提亚雷餐厅夜景
3. 莲花餐厅夜景

1. Lobby
2. Dining hall in Lotus
3. Dining hall in Hibiscus

1. 大堂
2. 莲花餐厅
3. 木槿餐厅

1. Salon Anteres suite
2. Over-water bungalow bedroom
3. Guestroom panoramic
4. Bathroom

1. 安特里斯套房客厅
2. 水上屋卧室
3. 全景房
4. 浴室

Hilton Bora Bora Nui
Resort & Spa

希尔顿波拉波拉岛度假村

Completion date: 2009
Location: Bora Bora, French Polynesia
Designer: Cabinet Tropical Architecture
de Pierre Lacombe; The Presidential Villas designed by
Philippe Villeroux
Photographer: Hilton Bora Bora Nui Resort & Spa
Area: 80,000 sqm

竣工时间：2009年
项目地点：法属波利尼西亚，波拉波拉岛
设计师：热带建筑；皮埃尔·拉康姆；总统别墅：
菲利普·维利洛克斯
摄影师：希尔顿波拉波拉岛度假村
项目面积：80,000平方米

The resort is located on the unique and exclusive island of Bora Bora, the famous "Pearl of the South Seas". The property combines the ultimate in luxury, authenticity and latest technology. All accommodations are in suites and villas with spectacular lagoon and ocean views. A stunning 1,000-square-metre infinity pool melts into views of the lagoon.

The hilltop spa features a spectacular 360° view of Bora Bora's famous Mt Otemanu and the neighbouring islands of Taha'a & Raiatea. The resort enjoys the exclusive use of its own private islet – Motu Tapu – for picnics and special events, wedding ceremonies and romantic dinners,

just five minutes away by boat.

The 122 luxury suites and villas are located on 80,000 square metres of lush and terraced land overlooking a private cove of crystalline waters. The place of Motu Toopua in sacred Polynesian mythology is reflected in authentic artefacts, decorations, and landscaping. The indigenous feel is carried into the luxurious oversized interiors of the suites and villas whose décor features marble tiles, exotic woods, and precious tapa cloths.

These suites and villas include 16 Lagoon-View Suites, especially convenient for families, 11 Hillside Villas with breathtaking panoramic views, 9 Garden Villas with exceptional garden-style accommodations, 38 Over-Water Villas with magnificent views of the island and lagoon, 44 Over-Water Deluxe Villas, with the true spirit of French Polynesia and endless views over the Pacific Ocean, and 2 Royal Over-Water Villas with unique panoramic location with a Jacuzzi on the deck. Two Presidential Over-Water Villas on two levels include well-being centre, outside swimming pool and jacuzzi.

1. Pool

2. Tamure grill

3. Iriatai Restaurant

4. Boutiques

5. Spa & fitness

6. Lagoon-view Suites

7. Villas

8. Reception

1. 游泳池

2. 塔西提烤肉店

3. 伊利亚太餐厅

4. 精品店

5. 水疗和健身中心

6. 湖景套房

7. 别墅

8. 前台接待处

1. Over-Water reception
2. Over-Water Villas
3. Resort plan
4. Exterior view of Presidential Over-Water Villas

1. 水上接待处
2. 水上别墅
3. 度假村平面图
4. 总统水上别墅外景

度假村坐落在独一无二的波拉波拉岛——著名的南海明珠。度假村汇集了极致奢华、真实和最新的技术。所有套房和别墅都享有壮丽的环礁湖和海洋美景。1,000平方米的无边界泳池与环礁湖的风景融为一体。

山顶温泉别墅享有波拉波拉岛上欧特马努山和塔哈瑞亚提亚岛360度的美景。度假村拥有一个专属私人小岛——塔普岛，专为野餐、特殊活动、婚礼和浪漫晚宴设计，乘船到达仅需5分钟。

122套奢华的套房和别墅坐落在80,000平方米的茂密梯田之上，俯瞰着私人海湾清澈的海水。图普阿岛神秘的波利尼西亚风情反映在艺术品、装饰品和景观设计之中。本土特色被引入了奢华的室内设计中。套房和别墅的设计以大理石饰面砖、异国森林和珍贵的塔帕织物为特色。

这些套房和别墅包括16套湖景套房（特别适合家庭居住）、11套山坡别墅（享有海洋全景）、9套花园别墅（非凡的花园住宿服务）、38套水上别墅（享有岛屿和湖泊的壮丽景色）、44套水上奢华别墅（真正的法属波利尼西亚风情，享有太平洋无尽的美景）、2套皇家水上别墅（平台上配有极可意按摩浴缸）。2套总统水上双层别墅内设有健身中心、露天游泳池和极可意按摩浴缸。

1. Iriatai Restaurant
2. Upa Upa bar
3. Over-water reception desk

1. 伊利亚太餐厅
2. 乌帕乌帕酒吧
3. 水上接待处

1. Royal Over-Water Villa
2. Hina spa
3. Presidential Over-Water Villa's well-being room

1. 皇家水上别墅
2. 希娜水疗中心
3. 总统水上别墅美容室

1. Presidential Over-Water Villa Twin bedroom
2. Presidential Over-Water Villa King bedroom
3. Royal Over-Water Villa bedroom

1. 总统水上别墅双人间
2. 总统水上别墅国王卧室
3. 皇家水上别墅卧室

Hilton Moorea Lagoon Resort & Spa

希尔顿茉莉雅岛潟湖度假村

Completion date: 2009
Location: Moorea, French Polynesia
Designer: Cabinet IHII de Grandou
Photographer: Hilton Moorea Lagoon Resort & Spa
Area: 10,000 sqm

竣工时间：2009年
项目地点：法属波利尼西亚，茉莉雅岛
设计师：格朗多IHII建筑事务所
摄影师：希尔顿茉莉雅岛泻湖度假村
项目面积：10,000平方米

Heart-shaped Moorea is also called the "Island of Love" and is said to be the inspiration for James Michener's mythical Bali Hai and the backdrop for the paintings of French impressionist, Paul Gauguin. Situated just 17 kilometres across the Sea of Moons from Papeete, this pretty sibling is a verdant green paradise.

All accommodations are in bungalows. The resort offers every modern convenience set in a luxurious décor of Polynesian inspiration. A white sand beach and lush tropical gardens are your surroundings while unwinding in the palm tree-lined infinity pool. An over-water bar and walkway are only a few steps away. Or consider the

myriad of ways to enjoy the turquoise lagoon: diving, snorkeling, canoeing, sailing…

There are totally 104 thatched-roof spacious bungalows: 25 Over-Water Bungalows, 29 Panoramic Over-Water Bungalows, 19 Garden Bungalows with pool, 26 Deluxe Garden Bungalows with pool, 3 Lagoon Bungalows and 2 Garden Pool Suites.

The Over-Water Bungalows have a fully-equipped private sun deck (lounge chairs, a table and chairs) with direct access to the lagoon. The Over-Water Bungalows have glass windows located on the floor for viewing the ballet of tropical fish below. The Garden and Deluxe Garden Bungalows offer majestic views from your private terrace or outdoor shower. The infinity free-form pool with its Jacuzzi overlooking the beach and the lagoon is only minutes away.

The beach-level, teak-accented spa is an oasis of relaxation. Polynesian spa attendants prepare scented baths infused with a choice of essential oils and filled with bright red hibiscus and white tiare flowers. Local products such as coconut monoi oil are used for the resort's world-class spa treatments.

心形的茉莉雅岛又被称作"爱之岛"，据说是詹姆斯·米切尔的神话《巴厘海》和法国印象派画家保罗·高更的绘画作品的原型。度假村距离首都帕皮提仅17千米，二者中间隔着月亮海。美丽的茉莉雅岛简直是郁郁葱葱的绿色天堂。

度假村采用别墅的形式，在奢华的波利尼西亚风情之中添加了现代便利设施。在棕榈树包围中的无边际泳池四周是洁白的沙滩和茂密的热带花园。水上酒吧和木板路距离别墅仅几步之遥。此外，人们还可以考虑大量的湖边活动：潜水、潜泳、独木舟、帆船……

度假村共有104座茅草屋顶别墅，其中包括：25套水上别墅、29套全景水上别墅、19套花园泳池别墅、26套奢华花园泳池别墅、3套湖景别墅、2套花园泳池套房。

水上别墅拥有一个设施齐全的私人日光浴平台，配有休息椅和桌椅，直接通往环礁湖。人们可以透过水上别墅的玻璃窗欣赏下方热带鱼的水中芭蕾表演。花园别墅和豪华花园别墅的私人平台或露天淋浴展现了绝妙的景色。无边际泳池和极可意按摩浴缸远眺着沙滩，距离环礁湖仅几步之遥。

海滩水疗馆以柚木为主要装饰，是放松休闲的绿洲。波利尼西亚水疗侍者将为宾客准备添加了精油的香薰洗浴，水中撒满了鲜红的木槿花和白栀子花。椰子精油等本土产品将被使用在度假村的顶级温泉水疗之中。

1. Infinity pool
2. Garden Pool Suite pool deck
3. Rotui bar
4. Resort plan

1. 无边界游泳池
2. 花园泳池套房的泳池平台
3. Rotui酒吧
4. 度假村平面图

1. Lobby
2. Fish pond
3. Restaurant "Arii Vahine"
4. Grill & Bar "Rotui"
5. Bar "Eineo"
6. Bar "Toatea"
7. Swimming pool
8. Tennis
9. Fitness centre
10. Spa
11. Nautic centre
12. Shop
13. Conference room
14. Parking
15. Pontoon
16. Heliport
17. Activities desk
18. Diving centre
19. Panoramic Over-Water Bungalow
20. Over-Water Bungalow
21. Deluxe Garden Bungalow with pool
22. Garden Bungalows with pool
23. Garden Pool Suite

1. 大堂
2. 鱼塘
3. 阿里瓦茵餐厅
4. 罗图烧烤酒吧
5. 艾尼欧酒吧
6. 托亚蒂酒吧
7. 游泳池
8. 网球场
9. 健身中心
10. 水疗中心
11. 水手中心
12. 商店
13. 会议室
14. 停车场
15. 浮桥
16. 直升机停机坪
17. 活动报名柜台
18. 潜水中心
19. 全景水上别墅
20. 水上别墅
21. 豪华花园泳池别墅
22. 花园泳池别墅
23. 花园泳池套房

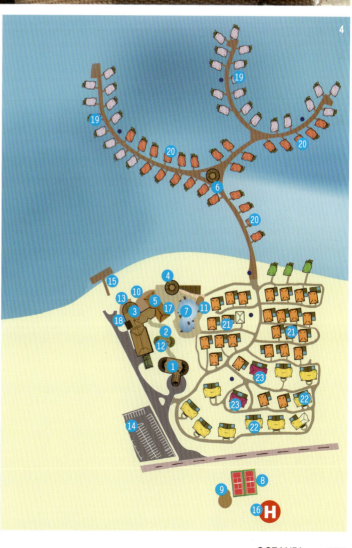

1. Garden Pool Suite
2. Meeting room
3. Spa room

1. 花园泳池套房
2. 会议室
3. 水疗室

1. Garden Pool Suite living room
2. Over-Water Bungalow bedroom
3. Over-Water Bungalow bedroom

1. 花园泳池套房客厅
2. 水上别墅卧室
3. 水上别墅卧室

Index 索引